# GEOProcessing 2020

The Twelfth International Conference on Advanced Geographic Information Systems, Applications, and Services

November 21 – 25, 2020

Valencia, Spain

**GEOProcessing 2020 Editors**

Claus-Peter Rückemann, Westfälische Wilhelms-Universität Münster (WWU) / DIMF / Leibniz Universität Hannover, Germany

Yerach Doytsher, Technion - Israel Institute of Technology, Haifa, Israel

Thomas Ritz, FH Aachen, Germany

**Printed from e-media with permission by:**

Curran Associates, Inc.
57 Morehouse Lane
Red Hook, NY 12571



International Academy, Research, and Industry Association (IARIA)
412 Derby Way
Wilmington, DE 19810

Phone: (408) 893-6407
Fax: (408) 527-6351

petre@iaria.org

**Additional copies of this publication are available from:**

Curran Associates, Inc.
57 Morehouse Lane
Red Hook, NY 12571 USA
Phone: 845-758-0400
Fax: 845-758-2633
Email: curran@proceedings.com
Web: www.proceedings.com

## Table of Contents

# GEOProcessing 2020

# Forward

The Twelfth International Conference on Advanced Geographic Information Systems, Applications, and Services (GEOProcessing 2020) addressed the aspects of managing geographical information and web services.

The goal of the GEOProcessing 2020 conference was to bring together researchers from the academia and practitioners from the industry in order to address fundamentals of advances in geographic information systems and the new applications related to them using the Web Services. Such systems can be used for assessment, modeling and prognosis of emergencies

GEOProcessing 2020 provided a forum where researchers were able to present recent research results and new research problems and directions related to them. The topics covered aspects from fundamentals to more specialized topics such as 2D & 3D information visualization, web services and geospatial systems, geoinformation processing, and spatial data infrastructure.

We take this opportunity to thank all the members of the GEOProcessing 2020 Technical Program Committee as well as the numerous reviewers. The creation of such a broad and high-quality conference program would not have been possible without their involvement. We also kindly thank all the authors who dedicated much of their time and efforts to contribute to the GEOProcessing 2020. We truly believe that, thanks to all these efforts, the final conference program consists of top quality contributions.

This event could also not have been a reality without the support of many individuals, organizations, and sponsors. We are grateful to the members of the GEOProcessing 2020 organizing committee for their help in handling the logistics and for their work to make this professional meeting a success.

We hope that GEOProcessing 2020 was a successful international forum for the exchange of ideas and results between academia and industry and for the promotion of progress in geographic information research.

**GEOProcessing 2020 Chairs**

**GEOProcessing 2020 General Chair**

Claus-Peter Rückemann, Westfälische Wilhelms-Universität Münster (WWU) / DIMF / Leibniz Universität Hannover, Germany

**GEOProcessing Steering Committee**
Thomas Ritz, FH Aachen, Germany
Yerach Doytsher, Technion - Israel Institute of Technology, Haifa, Israel

**GEOProcessing 2020 Advisory Committee**
Monica De Martino, Consiglio Nazionale delle Ricerche - Genova, Italy
Jianhong Cecilia Xia, Curtin University, Australia

**GEOProcessing 2020 Publicity Chair**
Lorena Parra, Universitat Politecnica de Valencia, Spain

# GEOProcessing 2020
# Committee

## GEOProcessing 2020 General Chair

Claus-Peter Rückemann, Westfälische Wilhelms-Universität Münster (WWU) / DIMF / Leibniz Universität Hannover, Germany

## GEOProcessing 2020 Steering Committee

Thomas Ritz, FH Aachen, Germany
Yerach Doytsher, Technion - Israel Institute of Technology, Haifa, Israel

## GEOProcessing 2020 Advisory Committee

Monica De Martino, Consiglio Nazionale delle Ricerche - Genova, Italy
Jianhong Cecilia Xia, Curtin University, Australia

## GEOProcessing 2020 Publicity Chair

Lorena Parra, Universitat Politecnica de Valencia, Spain

## GEOProcessing 2020 Technical Program Committee

Alia I. Abdelmoty, Cardiff University, Wales, UK
Danial Aghajarian, Georgia State University, USA
Nuhcan Akçit, Middle East Technical University, Turkey
Zaher Al Aghbari, University of Sharjah, UAE
Heba Aly, University of Maryland, College Park, USA
Francisco Javier Ariza López, Escuela Politécnica de Jaén - Universidad de Jaén, Spain
Thierry Badard, Centre de Recherche en Géomatique (CRG) | Université Laval, Canada
Abderrazak Bannari, Arabian Gulf University, Bahrain
Fabian Barbato, Ort University of Uruguay, Uruguay
Melih Basaraner, Yildiz Technical University, Turkey
Peter Baumann, rasdaman GmbH Bremen / Jacobs University Bremen, Germany
Mete Celik, Erciyes University, Turkey
Dickson K.W. Chiu, The University of Hong Kong, Hong Kong
Keith C. Clarke, University of California, Santa Barbara, USA
Alexandre Corrêa da Silva, HEX Geospatial Technologies, Brazil
Monica De Martino, CNR-IMATI (National research Council, Institute of applied Mathematics and Information technology), Italy
Cláudio de Souza Baptista, University of Campina Grande, Brazil
Yerach Doytsher, Technion - Israel Institute of Technology, Haifa, Israel
Suzana Dragicevic, Simon Fraser University, Canada
Emre Eftelioglu, Cargill Inc., USA
Süleyman Eken, Kocaeli University, Turkey
Salah Er-Raki, Université Cadi Ayyad, Morocco
Javier Estornell, Universitat Politècnica de València, Spain

Jamal Ezzahar, Université Cadi Ayyad, Morocco
Francisco R. Feito, University of Jaén, Spain
Anabella Ferral, Instituto de Altos Estudios Espaciales Mario Gulich | Centro Espacial Teófilo Tabanera - CONAE, Córdoba, Argentina
Douglas Galarus, Utah State University, USA
Erica Goto, University of California Santa Barbara (UCSB), USA
William Grosky, University of Michigan-Dearborn, USA
Mohd Helmy Abd Wahab, Universiti Tun Hussein Onm Malaysia, Malaysia
Arif Hidayat, Monash University, Australia / Brawijaya University, Indonesia
Masaharu Hirota, Okayama University of Science, Japan
Qunying Huang, University of Wisconsin, Madison, USA
Chih-Cheng Hung, Kennesaw State University - Marietta Campus, USA
Sergio Ilarri, University of Zaragoza, Spain
Katerina Kabassi, Ionian University, Greece
Antonios Karatzoglou, Robert Bosch GmbH, Germany
Hassan A. Karimi, University of Pittsburgh, USA
Baris M. Kazar, Oracle America Inc., USA
Saïd Khabba, Université Cadi Ayyad, Marrakech, Morocco
Kyoung-Sook Kim, National Institute of Advanced Industrial Science and Technology (AIST), Tokyo, Japan
Mel Krokos, University of Portsmouth, UK
Piyush Kumar, Florida State University, USA
Robert Laurini, INSA Lyon | University of Lyon, France
Dan Lee, Esri Inc.,USA
Lassi Lehto, Finnish Geospatial Research Institute, Finland
Xinghua Li, Wuhan University, China
Jugurta Lisboa-Filho, Federal University of Viçosa, Brazil
Ying Lu, DiDi Research America, Mountain View, USA
Ahmed Mahmood, Google, USA
Dipankar Mandal, Indian Institute of Technology Bombay, India
Ali Mansourian, Lund University, Sweden
Jesús Martí Gavilá, Research Institute for Integrated Management of Coastal Areas (IGIC) | Universitat Politècnica de València, Spain
Sara Migliorini, University of Verona, Italy
Sobhan Moosavi, The Ohio State University, USA
Tathagata Mukherjee, The University of Alabama in Huntsville, USA
Beniamino Murgante, University of Basilicata, Italy
Ahmed Mustafa, The New School University, New York, USA
Aldo Napoli, MINES ParisTech - CRC, France
Maurizio Napolitano, Fondazione Bruno Kessler, Trento, Italy
Javier Nogueras-Iso, University of Zaragoza, Spain
Alexey Noskov, Philipps University of Marburg, Germany
Xiao Pan, Shijiazhuang Tiedao University, China
Shray Pathak, School of Geographic Sciences | East China Normal University, Shanghai, China

Kostas Patroumpas, Athena Research Center, Greece
Davod Poreh, Università degli Studi di Napoli "Federico II", Italy
Satish Puri, Marquette University, Wisconsin, USA
Thomas Ritz, FH Aachen, Germany
Ricardo Rodrigues Ciferri, Federal University of São Carlos (UFSCar), Brazil
Armanda Rodrigues, Universidade NOVA de Lisboa | NOVA LINCS, Portugal
Claus-Peter Rückemann, Westfälische Wilhelms-Universität Münster (WWU) / DIMF / Leibniz Universität Hannover, Germany, Germany
Ibrahim Sabek, University of Minnesota, USA
André Sabino, Universidade Autónoma de Lisboa, Portugal
Markus Schneider, University of Florida, USA
Raja Sengupta, McGill University, Montreal, Canada
Shih-Lung Shaw, University of Tennessee, Knoxville, USA
Yosio E. Shimabukuro, Brazilian Institute for Space Research - INPE, Brazil
Spiros Skiadopoulos, University of the Peloponnese, Greece
Dimitris Skoutas, Athena Research Center, Greece
Francesco Soldovieri, Istituto per il Rilevamento Elettromagnetico dell'Ambiente - Consiglio Nazionale delle Ricerche (CNR), Italy
Katia Stankov, University of British Columbia, Canada
Payam Tabrizian, IDEO, San Fransisco, USA
Ergin Tari, Istanbul Technical University, Turkey
Brittany Terese Fasy, Montana State University, USA
Roger Tilley, University of California, Santa Cruz, USA
Goce Trajcevski, Iowa State University, USA
Linh Truong-Hong, Delft University of Technology, Netherlands
Taketoshi Ushiama, Kyushu University, Japan
Marlène Villanova-Oliver, Univ. Grenoble Alpes - Grenoble Informatics Lab, France
Tin Vu, University of California, Riverside, USA
Hong Wei, University of Maryland, College Park, USA
John P. Wilson, University of Southern California, USA
Jianhong Cecilia Xia, Curtin University, Australia
Ningchuan Xiao, The Ohio State University, USA
Xiaojun Yang, Florida State University, USA
Qiangqiang Yuan, School of Geodesy and Geomatics | Wuhan University, China
F. Javier Zarazaga-Soria, University of Zaragoza, Spain
Shenglin Zhao, Tencent, Shenzhen, China
Qiang Zhu, University of Michigan - Dearborn, USA

**Copyright Information**

# Taming Near Repeat Calculation for Crime Analysis via Cohesive Subgraph Computing

Zhaoming Yin

Open Data Processing Platform Team,
Alibaba Cloud
Hangzhou, Zhejiang, China
Email: stplaydog@gmail.com

Xuan Shi

Department of Geosciences
University of Arkansas
Fayetteville, Washington County, USA
Email: xuanshi@uark.edu

***Abstract*—Near Repeat (NR) is a well-known phenomenon in crime analysis, assuming that crime events exhibit correlations within a given time and space frame. Traditional NR calculation would generate two event pairs if two events happened within a given space and time limit. When the number of events is significant, however, NR calculation is time consuming and how these pairs are organized has not yet been explored. In this paper, we designed a new approach to calculate clusters of NR events efficiently. To begin with, R-tree is utilized to index crime events. A single event is represented by a vertex, whereas edges are constructed by range-querying the vertex in R-tree; this way, a graph is formed. Cohesive subgraph approaches are applied to identify the event chains. k-clique, k-truss, k-core plus Density-based Spatial Clustering of Applications with Noise (DBSCAN) algorithms are implemented in sequence to their varied range of abilities to find cohesive subgraphs. Real-world crime data in Chicago, New York, and Washington DC are utilized to conduct experiments. The experiments confirmed that near repeat has a substantial effect on real big crime data by conducting Map-reduce empowered Knox tests. The performances of 4 different algorithms are validated, with the quality gauged by the distribution of the number of cohesive subgraphs and their clustering coefficients. The proposed framework is the first to process the real crime data of million records and is the first to detect NR events with a size of more than 2.**

***Keywords*–*Near-Repeat; Graph Analysis.***

## I. INTRODUCTION

In criminal research, it was found that when a crime incident takes place at a given geographical location, its neighboring areas would have a higher possibility of experiencing follow-up incidents in a short period [16] [19] [25]. When the first incident occurs at a specific time, the follow-up incident at the same location and close to the initial time is a repeat. The incidents that occur near the space and time of the initiator are called near-repeat. Such repeat and near-repeat phenomena have been found from burglaries and gun violence studies, and have important implications to dispatch police force in crime mitigation activities [16] [21] [25]. To prove the near-repeat effect, the classic way is to use the Knox test method [16] [19] [22]. The general idea of Knox test is to calculate the pairwise distance (in terms of space and time) between different crime events, and place the event pair into different bins of a table; the residual value of each specific entry of the table is calculated to indicate how random these pairs are organized into the range. The issue with this method is that its time complexity is $O(n^2)$ ($n$ is the number of crime events). When dealing with big real-world data, it will take days, if not months, to finish the computing task.

When near-repeat research only considers the space-time interaction among every two incidents, a complete space-time event chain is more appropriate to differentiate such a scenario of separate space-time pairs [21]. For example, when three shooting events (A and B), (B and C), (A and C) comply with the near-repeat definition, a three-event chain can be identified. In this case, it would be more meaningful to identify the correlation between multiple incidents rather than just two. Event chain analysis improves our understanding of the role of space and time among series of shooting events, or other types of crime events. The significant existence of paired shooting events does not mean the meaningful presence of multiple shooting event chains in the same space-time context. Otherwise, all initiators or follow-up shooting events should be close to each other and form a spatial cluster in a city.

Enumerating event chains in a brute-force way would be extremely difficult because the time complexity grows exponentially. Nevertheless, we can abstract the problems of near-repeat event chain detection by dividing it into two separate issues: 1) detecting near repeat pairs efficiently; 2) clustering or chaining near repeat pairs with high speed.

To begin with, the most efficient way to avoid unnecessary pairwise computation of each crime event is to use an index to organize events, such that one only needs to query its spatial-temporal adjacent events to generate event pairs, and R-tree [12] is the right choice. Once all event pairs are detected, they can be represented as a graph.

Furthermore, detecting a chain of near-repeat events can be modeled as a cohesive subgraph enumeration problem [9]. Ideally, all events should have connections between each other within a cohesive subgraph, and such a subgraph is a k-clique ($k$ is the number of vertices in the subgraph) [15]; Nevertheless, in the real world, the graph is massive, and approximation methods are more efficient than exact algorithms [18]. Rather than asking all vertices in a subgraph to have a connection between each other, a k-core [5] only asks that each vertex in the subgraph has k degrees, and this restriction is relaxed. A variant of Density-based Spatial Clustering of Applications with Noise (DBSCAN) algorithm [6] can be applied to detect clusters of spatial-temporal data. The DBSCAN algorithm is fast with a complexity of $O(Vlog(V))$. One of the alternative versions requires that each edge in the subgraph should be in $k-2$ triangles; this is called k-truss [9]. In recent years, lots of advances had been made in the area of truss decomposition, regarding speed [23], a variance of graph [14], and data streaming [13], which make the k-truss algorithm a preferred alternative to the k-clique method. All three alternatives to the k-clique algorithm have polynomial complexity.

**Organization.** In this paper, we will discuss a framework to incorporate the methods of indexing crime events, computing event pairs, and detecting near-repeat event chains applying k-clique, k-core, DBSCAN, and k-truss algorithms separately. Section 2 formally defines the near repeat chain detection problem, gives the necessary notations, and describes the framework of our near repeat chain detection methods; Section 3 reports the experimental results using real-world data; the Conclusion is derived in the last section of this paper.

## II. Event Chain Calculation through Graph Analytics

### A. *Using graph to represent Crime Events*

Suppose we use a vertex $v$ to represent an event that occurred in location $(x, y)$ and at time $t$. Moreover, if two vertices $v_1$ and $v_2$ representing different events occurred within a given time and space constraint, an undirected edge $(v_1, v_2)$ will be used to connect them. If there are $V$ number of events and $E$ number of event pairs, the resulting vertices and edges form a graph $G$ (In this article, we assume that there is only one undirected edge between any two vertices). If there exists a set of events with $n$ number of events, each event is paired with every other $n-1$ event. We call this set of events an event chain. A subgraph $g$ in $G$ can represent this event chain such that each vertex in the $g$ will have edges connecting every other vertex in $g$. Figure 2(a) shows an example of 8 crime events and 13 event pairs. In the figure, subgraph induced from vertices $\{1, 2, 3\}$ shows an example of a 3-event chain which is a triangle. Furthermore, subgraph induced from vertices $\{0, 1, 3, 4\}$ shows an example of a 4-event chain, which is a 4-clique. Press ENTER or type command to continue The degree of a vertex $v$ is defined as the number of edges connecting $v$. Take Figure 2(a) for example, the degree for vertex 1 is 5, and the degree for vertex 5 is 2. The definition of k-clique [7] is, each vertex in k-clique has degree of exactly $k-1$; Figure 2(b) shows a 4-clique subgraph. Similarly, k-core [5] is the subgraph in which each of its vertices has a degree of no less than $k$. Figure 2(c) shows a 3-core subgraph, k-DBSCAN [6] is also a degree based cohesive subgraph, the method leverages k-degree vertices to greedily expand clusters(we will discuss the detail later), and Figure 2(a) itself is the subgraph induced by 3-DBSCAN. As for an edge $e$ in $G$, the number of triangles it belongs to is called the support. For instance, the support for edge $(1, 2)$ is 3, and the support for edge $(2, 7)$ is 1. k-truss [23] is the subgraph with each of its edges having support no less than $k-2$; Figure 2(d) shows a 3-truss induced from the original graph. The clustering coefficient [24] evaluates the tightness of the connection in a cohesive subgraph. If we use $coe(g)$ to denote a graph's clustering coefficient, empirically we have $coe(g_{k-clique}) \geq coe(g_{k-truss}) \geq coe(g_{k-core})$ [23].

### B. *Near Repeat Event Chain Detection Algorithm*

*1) Algorithm description:* Given a set of crime events, we can represent each event using a coordinate $x$, $y$, and a time $t$ of crime type $p$. We would transform the coordinate using UTM format [8]. The process of finding near repeat crime event chain can be formulated as the following two steps:

- Create a graph based on the spatial-temporal coordinates of a specific crime type; Since computing all pairs of events is expensive, we will build an R-tree [12] using 3-dimensional coordinates x, y, and t. A vertex forms edges with its neighbors by specifying some query criteria in R-tree.
- Based on the graph created at step 1, compute the cohesive subgraphs such as k-clique, k-core, k-DBSCAN, or k-truss. Optimization methods might be applied; for instance, we can divide the graph into small graphs, if multiple connected components are detected [10].

The algorithm is described in Figure 1. The complexity of the algorithm could be divided into two steps. Suppose we have $V$ events, and $E$ event pairs. The complexity for the first graph generation process is dependent on data, and can not guarantee a worst-case complexity, but its lower bound is $O(V)$. The complexity of computing k-clique is NP-Hard [7], k-core is $O(E)$ [5], k-DBSCAN is $O(Vlog(V))$ [11], and k-truss is $O(E^{1.5})$ [23].

*2) k-clique enumeration Revisited:* The maximum clique problem is a widely researched area, and there are lots of papers on this topic since the general algorithmic framework for clique enumeration algorithm is different from the other three algorithms, we will not spend too much effort on this theme.

*3) k-core Computation Revisited:* Figure 3 displays the skeleton of the k-core algorithm. Vertices are sorted by their degrees in ascending order, and the criteria of $k$ start from 3. Vertices with degree less than $k$ and their adjacent edges are removed from the graph $G$ and the neighbor vertices of these removed vertices (we use $nb(v)$ to denote neighbors of $v$) will update their degrees accordingly. Once there is no such vertex to be removed, the remaining graph will be placed in the k-core class $T_k$, and $k$ will be incremented, and the removing procedure will start again. The procedure continues until there is no vertex to be removed.

*4) k-DBSCAN Computation Revisited:* k-DBSCAN is a density-based clustering algorithm. In this paper, we translate the k-DBSCAN algorithm into the equivalence of the cohesive

**Input**: Set of crime events $C$, range criteria ($r_x$, $r_y$, $r_t$)
**Output**: Set of near repeat event chains, $T$

Initialize R-tree $R$ ;
**for** $v$ *in* $C$ **do**
    $R.insert(v)$ ;
**end**
**for** $v$ *in* $C$ **do**
    ($x$, $y$, $t$) = coordinate of $v$ ;
    $S$ = R.retrieve([$x \pm r_x$, $y \pm r_y$, $t \pm r_t$]) ;
    **for** $u$ *in* $S$ **do**
        add edge ($u$, $v$) to $G$ ;
    **end**
**end**
$T$ = cohesive_subgraph_algo($G$) ;
**return** $T$ ;

Figure 1. Near repeat event chain calculation.

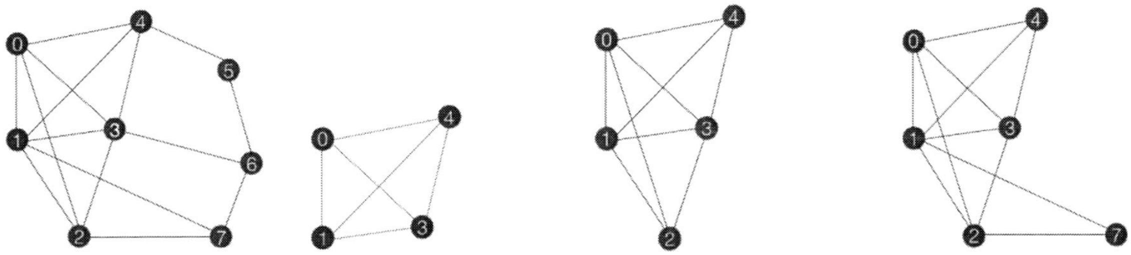

(a) Origin (b) 4-clique (c) 3-core (d) 3-truss

Figure 2. Example using graph to represent crime event pairs/chains.

subgraph algorithm. The algorithm is as Figure 4 shows. The algorithm finds a vertex $v$ with $deg(v) \geq k$ from $k = 3$, then expand it (as Figure 5 shows), every expansion will result in the vertices in the expansion being marked as visited. If there is no vertex to be expanded, we will remove all the vertices that are not visited from $G$. The procedure continues until there is no vertex to be removed.

*5) k-truss Decomposition Revisited:* Truss decomposition is firstly introduced in paper [9] to detect possible subgroups within a social network. It is pretty useful in community detection. New and efficient algorithms are introduced to compute truss efficiently [23]. The idea of the algorithm is to compute the support for each edge first. For each edge, the $O(d)$ complexity algorithm for triangle enumeration will be applied, $d$ is the larger degree of the two vertices forming an edge. In this paper, we will use Compressed Sparse Row (CSR) [17] to store the edge array. The skeleton of the k-truss algorithm is shown in Figure 6. Then the edges are sorted

```
Input: Graph G
Output: k-core (k ≥ 3) T
1  k = 3, T_k = ∅ ;
2  Compute deg(v) for v ∈ G ;
3  Sort all the vertices in ascending order by degree and
   place them in U ;
4  while ∃v in U such that deg(v) < k do
5      W = nb(v) ;
6      for w in W do
7          deg(w) − − ;
8          deg(w) − − ;
9          Reorder w in U according to its new degree;
10     end
11     Remove v and its adjacent edges from G ;
12     Remove v from U ;
13 end
14 if Not all v in U are removed then
15     T[k] = G ;
16     k + + ;
17     goto step 4 ;
18 end
19 return T ;
```

Figure 3. k-core computation.

```
Input: Graph G
Output: k-DBSCAN (k ≥ 3), T
1  k = 3, T_k = ∅ ;
2  compute deg(v) for v ∈ G ;
3  Sort all the vertices in descending order by degree and
   place them in U ;
4  while ∃v such that deg(v) ≥ k and
   visited(v) == false do
5      expand(G, v, k) ;
6  end
7  while ∃v such that visited(v) == false do
8      Remove v and its adjacent edges from G ;
9      Remove v from U ;
10 end
11 if Not all v in U are removed then
12     T[k] = G ;
13     k + + ;
14     set all v in U as unvisited ;
15     goto step 4 ;
16 end
17 return T ;
```

Figure 4. k-DBSCAN computation.

```
Input: Graph G, vertex v, k
1 W = nb(v) ;
2 for w in W do
3     if visited(w) == false then
4         visited(w) = true ;
5         if deg(w) ≥ k then
6             expand(w) ;
7         end
8     end
9 end
```

Figure 5. Expand procedure.

in the ascending order by their support. To compute the k-truss, every edge with support less than $k-2$, along with its incident vertices, will be removed. Moreover, the incident edges will update their support and their position in the edge array following similar methods k-core computation. Followed by the removal of all edges that do not form a k-truss, the remaining graph consists of k-trusses. Furthermore, the value of $k$ will be incremented, followed by the same edge removing steps until there are no edges left in the graph. In the algorithm, because the range of supports is already known, sorting can be done with $O(E)$ complexity using bucket sort.

## III. EXPERIMENTS

### A. Data Sets

The data used in this research contains the real crime data onto New York (NYC), Washington DC (DC), and Chicago (CHI) retrieved from data.gov [1] [2] [3]. The general information about the data is displayed in TABLE II. In the table, the granularity of time is in a day(s), and the granularity of space is in meters. #t means the number of crime types, #d means the number of duplicated events. #events are the number of events after combining duplications. We have removed the data that is not conformed to the right format (for instance, data that does not fall into the range in TABLE II ), and combined the duplicated entries ( for example, a crime of the same type happens at the same time of the same location, see TABLE II #d). In general, all three data sets have crime numbers of a million scale.

Since there are numerous crime types of DC and CHI data (see TABLE II #t), we only choose the crime types of burglary (BUR), robbery (ROB) and theft (TFT) for detailed discussion. We selected the spatial-temporal range limit $r_x = r_y = 100(meters)$ with $r_t = 10(days)$, and the feature of graphs generated applying this criteria have the property as TABLE I shows. In the table, #V is the number of vertices, #E is the number of edges, #CC is the number of connected components(we do not count the isolated vertices as CC), d_avg and d_var are the mean and variance of the diameters of connected components; c_avg and c_var are the mean and variance of clustering coefficient of connected components. We use Floyd Warshal [10] algorithm to calculate the all-pairs shortest path of each clique, and use this information to infer the diameter of each clique. As for the clustering coefficient [24], we use it to evaluate how densely these graphs are organized. In general, burglary and robbery are sparse near repeat events in comparison to theft concerning the number of vertices $\#V$, which is also indicated by a more substantial number of edges and connected components $\#CCand\#E$. The diameter and clustering coefficient feature also indicates that theft has large clusters, and these clusters are dense. Inferred from the table, the graphs in all data obey small-world property because the diameters of the graphs are small [24].

**Input**: Graph $G$
**Output**: k-truss $(k \geq 3)$, $T$
$k = 3$, $T_k = \emptyset$ ;
compute sup($e$) for $e \in G$ ;
Sort all the edges in ascending order of their support and place them in $U$ ;
**while** $\exists e$ *in* $U$ *such that* $sup(e) \leq (k-2)$ **do**
  $e = (u, v)$ with the lowest support ;
  $W = nb(u) \cap nb(v)$ ;
  **for** $w$ *in* $W$ **do**
    $sup(u, w) - -$ ;
    $sup(v, w) - -$ ;
    Reorder (u, w) and (v, w) according to their new support;
  **end**
  Remove $e$ from $G$ ;
  Remove $e$ from $U$ ;
**end**
**if** *Not all e in U are removed* **then**
  $T[k] = G$ ;
  $k + +$ ;
  goto step 4 ;
**end**
**return** $T$ ;

Figure 6. k-truss computation.

TABLE I. GENERAL INFORMATION OF GRAPHS.

| | #V | #E | #CC | d_avg | d_var | c_avg | c_var |
|---|---|---|---|---|---|---|---|
| | | | | NY | | | |
| BUR | 187k | 112k | 24k | 1.25 | 0.64 | 0.12 | 0.064 |
| ROB | 198k | 152k | 27k | 1.34 | 1.38 | 0.13 | 0.068 |
| TFT | 421k | 1.5m | 55k | 1.77 | 6.66 | 0.20 | 0.089 |
| | | | | DC | | | |
| BUR | 156k | 54k | 13k | 1.26 | 0.80 | 0.10 | 0.054 |
| ROB | 54k | 32k | 6k | 1.40 | 1.30 | 0.138 | 0.069 |
| TFT | 344k | 1.1m | 33k | 1.75 | 7.78 | 0.17 | 0.080 |
| | | | | CHI | | | |
| BUR | 197k | 118k | 29k | 1.29 | 0.56 | 0.12 | 0.060 |
| ROB | 124k | 68k | 14k | 1.34 | 0.96 | 0.11 | 0.058 |
| TFT | 650k | 3.4m | 89k | 1.84 | 7.95 | 0.19 | 0.081 |

### B. Knox test with Map-reduce

Firstly, we prove the existence of a near-repeat effect in a real big data set by conducting a Knox test on the data set. We implement the Knox test using the Map-reduce framework on Amazon AWS EMR and store the input/output on S3. The program is written in python and runs with Hadoop streaming [26] mode. For New York and Chicago theft data set, we used a cluster of 16 nodes, and for all the other data sets, we used a cluster of 4 nodes (for a budget reason).

The computational time is recorded and is shown on TABLE III. To the best of our knowledge, the previous Knox test research on crime data is orders of magnitudes smaller than our data. Since the complexity of the Knox test is $O(n^2)$, it is not practical to compare the timing of these results against the previous experiments. Hence in this paper, we claim that our method can finish the Knox test within a reasonable time from less than an hour to approximately 10 hours using big real-world data, which is not possible with the previous methods.

To construct the Knox test table [22], we have set the distance step as 100 meters and the time step as 14 days. The Knox test result is shown in Figure 7 using a heatmap. In the figure, we do not show the result of distance larger than $10 \times 100 = 1000$ meters and time difference larger than

TABLE II. GENERAL INFORMATION OF DATA.

| name | earliest | latest | min x | max x | min y | max y | #events | #t | #d |
|---|---|---|---|---|---|---|---|---|---|
| NY | 2006/06/04 | 2015/12/31 | 134239 | 1067186 | 121080 | 7220451 | 1123221 | 7 | 29k |
| DC | 1978/01/01 | 2015/12/31 | 4840550 | 18915876 | 777144 | 8480189 | 2130867 | 43 | 89k |
| CHI | 2001/01/01 | 2015/12/31 | 1092706 | 1205119 | 1813894 | 1951610 | 3102758 | 35 | 54k |

$4 \times 14 = 56$ days. It is evident from the heatmap that all three crime types in three cities exhibit near repeat effects because the upper left corner entries of each Knox test matrices have residual values that are significantly larger than other entries.

TABLE III. COMPUTATIONAL TIME (IN SECONDS) FOR KNOX TEST USING MAPREDUCE.

| City | BUR | ROB | TFT |
|---|---|---|---|
| NY | 13320 (4 nodes) | 13980 (4 nodes) | 13560 (16 nodes) |
| DC | 9240 (4 nodes) | 1440 (4 nodes) | 34320 (4 nodes) |
| CHI | 14340 (4 nodes) | 6000 (4 nodes) | 24060 (16 nodes) |

### C. *Near-repeat chain detection*

We implement the k-core, k-DBSCAN and k-truss algorithm using C++, and GCC compiler with the c++-11 features enabled; Furthermore, the code is freely available on GitHub with the package name OPTKIT. To build the spatial-temporal index with an R-tree package, we use the open-source implementation of [12]. As for the k-clique, and the graph properties, we use the boost graph library (BGL) [20]. The experiment was run on an AWS machine [4], with an m4.4xlarge Redhat instance. The instance has 16 cores, and each has a 2.4 GHz Intel Xeon E5-2676 v3 (Haswell) processor and 64 Gbs of memory, the disk size is 160 Gbs of EBS storage; the operating system is RHEL-7.2. Computational time is recorded by analyzing the log result using the glog library.

*1) Computational Time:* The computational time is divided into seven parts, including the time to 1) load the data, which includes parsing and reading spatial-temporal coordinates of CSV format. 2) build R-tree and edges based on querying the R-tree. 3) separate edges based on the connected component computation using BGL. 4-6) implement the k-truss, k-core, and k-DBSCAN algorithms. 7) calculate k-cliques.

The computational time results is an excerpt on TABLE IV, in the table, the load is the time for loading spatial-temporal information in CSV format. The R-tree is the time for building the R-tree index. The edges is the time to build edges based on R-tree, CC is the time to calculate connected components, the truss is the time to calculate k-truss, the core is the time to calculate k-core, the dbscan is the time to calculate k-DBSCAN. BGL is the time for the k-clique calculation using the boost graph library. No matter the size of the data, the dominant computational time is spent on the cohesive subgraph calculation. It is observed that when the graph is small and less dense, it takes less time to utilize BGL to compute graph properties. In case the graph size is expanding fast, with larger clusters, it takes a considerable amount of time to get the k-clique result. Consequently, the advantage of approximate cohesive subgraph computation will be distinct. It seems our

TABLE IV. RESULTS FOR COMPUTATIONAL TIME.

| | load | R-tree | edges | CC | truss | core | dbscan | BGL |
|---|---|---|---|---|---|---|---|---|
| | | | NY | | | | | |
| BUR | 0.54 | 0.50 | 0.15 | 0.11 | 6.15 | 1.53 | 4.30 | 2.14 |
| ROB | 0.62 | 0.51 | 0.21 | 0.13 | 7.05 | 1.97 | 4.16 | 2.00 |
| TFT | 1.24 | 1.10 | 1.06 | 0.80 | 10.55 | 4.05 | 4.71 | 302.79 |
| | | | DC | | | | | |
| BUR | 0.47 | 0.43 | 0.17 | 0.05 | 4.97 | 0.91 | 1.59 | 0.83 |
| ROB | 0.15 | 0.13 | 0.05 | 0.03 | 0.86 | 0.28 | 0.32 | 0.44 |
| TFT | 1.07 | 0.96 | 0.59 | 0.44 | 7.87 | 1.87 | 2.37 | 87.31 |
| | | | CHI | | | | | |
| BUR | 0.62 | 0.54 | 0.21 | 0.12 | 10.08 | 1.65 | 2.6 | 1.67 |
| ROB | 0.39 | 0.33 | 0.13 | 0.07 | 6.09 | 1.06 | 2.11 | 0.97 |
| TFT | 1.10 | 1.69 | 4.33 | 2.76 | 21.20 | 7.82 | 10.93 | 487.92 |

TABLE V. NUMBER OF K-CLIQUE DETECTED.

| | 3 | 4 | 5 | 6 | 7 | 8 | 9 | $\geq 10$ |
|---|---|---|---|---|---|---|---|---|
| | | | NY | | | | | |
| BUR | 4117 | 935 | 304 | 87 | 33 | 11 | 5 | 16 |
| ROB | 4867 | 1364 | 491 | 195 | 88 | 46 | 24 | 40 |
| | | | DC | | | | | |
| BUR | 1770 | 432 | 133 | 56 | 26 | 11 | 6 | 4 |
| ROB | 1073 | 314 | 118 | 45 | 20 | 12 | 8 | 8 |
| | | | CHI | | | | | |
| BUR | 4941 | 1039 | 223 | 41 | 10 | 5 | 1 | 0 |
| ROB | 2338 | 600 | 207 | 66 | 34 | 12 | 2 | 4 |

TABLE VI. NUMBER OF K-CORES DETECTED.

| | 3 | 4 | 5 | 6 | 7 | 8 | 9 | $\geq 10$ |
|---|---|---|---|---|---|---|---|---|
| | | | NY | | | | | |
| BUR | 1336 | 1295 | 803 | 307 | 163 | 60 | 35 | 94 |
| ROB | 1732 | 1944 | 1144 | 624 | 351 | 233 | 177 | 321 |
| TFT | 5227 | 7467 | 5787 | 4452 | 3124 | 2272 | 1824 | 8661 |
| | | | DC | | | | | |
| BUR | 641 | 664 | 306 | 193 | 114 | 66 | 52 | 34 |
| ROB | 415 | 496 | 285 | 159 | 87 | 42 | 33 | 68 |
| TFT | 2980 | 4090 | 2934 | 2415 | 1490 | 1153 | 951 | 6855 |
| | | | CHI | | | | | |
| BUR | 2092 | 1729 | 649 | 195 | 40 | 33 | 13 | 0 |
| ROB | 481 | 315 | 153 | 105 | 62 | 6 | 18 | 0 |
| TFT | 11223 | 12904 | 8549 | 6081 | 4329 | 3047 | 2234 | 15028 |

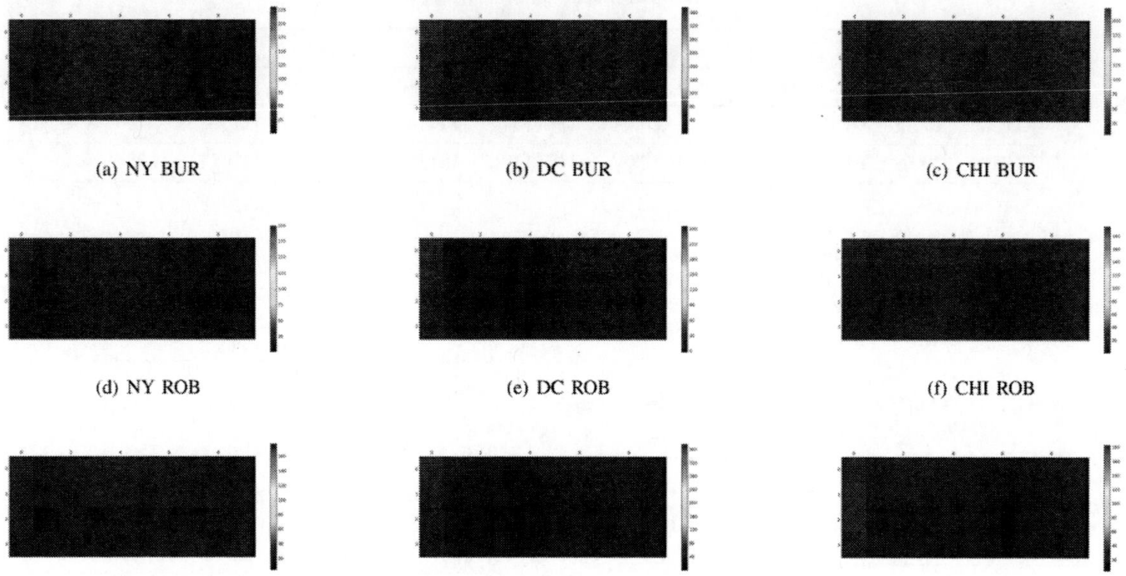

(a) NY BUR (b) DC BUR (c) CHI BUR

(d) NY ROB (e) DC ROB (f) CHI ROB

(g) NY TFT (h) DC TFT (i) CHI TFT

Figure 7. Knox test results. A $4 \times 10$ colored matrices represent each test, the row step is 14 days, and the column step is 100 meters. The residual value of that range plots the color of the entry.

TABLE VII. NNUMBER OF K-DBSCAN DETECTED.

| | 3 | 4 | 5 | 6 | 7 | 8 | 9 | $\geq 10$ |
|---|---|---|---|---|---|---|---|---|
| | | | NY | | | | | |
| BUR | 782 | 509 | 211 | 101 | 41 | 24 | 47 | 0 |
| ROB | 1 | 1068 | 707 | 389 | 243 | 148 | 110 | 224 |
| TFT | 1 | 3894 | 3391 | 2744 | 1967 | 1473 | 1187 | 5193 |
| | | | DC | | | | | |
| BUR | 1 | 362 | 194 | 124 | 77 | 45 | 31 | 29 |
| ROB | 415 | 496 | 285 | 159 | 87 | 42 | 33 | 68 |
| TFT | 2980 | 4090 | 2934 | 2415 | 1490 | 1153 | 951 | 6855 |
| | | | CHI | | | | | |
| BUR | 6 | 944 | 396 | 112 | 24 | 20 | 7 | 0 |
| ROB | 481 | 315 | 153 | 105 | 62 | 6 | 18 | 0 |
| TFT | 57 | 6071 | 4676 | 3547 | 2630 | 1859 | 1400 | 8608 |

TABLE VIII. NUMBER OF K-TRUSS DETECTED.

| | 3 | 4 | 5 | 6 | 7 | 8 | 9 | $\geq 10$ |
|---|---|---|---|---|---|---|---|---|
| | | | NY | | | | | |
| BUR | 4988 | 1217 | 401 | 120 | 48 | 16 | 8 | 69 |
| ROB | 6153 | 1885 | 738 | 340 | 166 | 87 | 48 | 109 |
| TFT | 20091 | 9628 | 5448 | 3420 | 2217 | 1538 | 1153 | 4412 |
| | | | DC | | | | | |
| BUR | 2178 | 600 | 206 | 94 | 43 | 17 | 8 | 6 |
| ROB | 1359 | 436 | 168 | 77 | 36 | 18 | 12 | 11 |
| TFT | 10564 | 5025 | 2871 | 1922 | 1349 | 1037 | 796 | 4180 |
| | | | CHI | | | | | |
| BUR | 5851 | 1227 | 253 | 51 | 12 | 5 | 1 | 0 |
| ROB | 2886 | 811 | 290 | 101 | 40 | 16 | 5 | 4 |
| TFT | 32345 | 14418 | 7815 | 4912 | 3299 | 2410 | 1885 | 9683 |

implementation is slower in the small graph case. Nevertheless, in general, the less cohesive the requirement of the results, the less time it takes to compute the result.

*2) Results comparison:* TABLES VI, VII, VIII, V show the distribution of the number of the cohesive subgraphs. Figure 8 shows the clustering coefficient of the cohesive subgraphs detected using different methods on different data. We exclude the results of k-clique because the clustering coefficient is always 1. Although there are some variations of the results, we can, in general, conclude that k-truss is better than the k-DBSCAN, which is better than the k-core method. It is also worth noting that in many results when it comes to the cohesive subgraphs of large k (approximately $k > 10$), subgraphs detected by the k-DBSCAN and k-core algorithm have very stable clustering coefficients, which might indicate some specific and stable graph patterns detected by these algorithms when $k$ is large.

## IV. CONCLUSION

In this paper, we have designed a Mapreduce based Knox test algorithm to help to prove the existence of a near-repeat effect on big data. We explore to identify efficient algorithms to derive near-repeat event chains. By representing crime events into a graph enabled by R-tree indexing, the near repeat crime chains can be derived through cohesive subgraph analysis. Four different cohesive subgraph analysis methods are implemented using AWS resources and compared concerning time and quality. The proposed solution has never been applied in the prior works on crime analysis and will have a broader impact on this research front in the future. However, there are still potential improvements to be made.

To begin with, we should perform the Knox test using

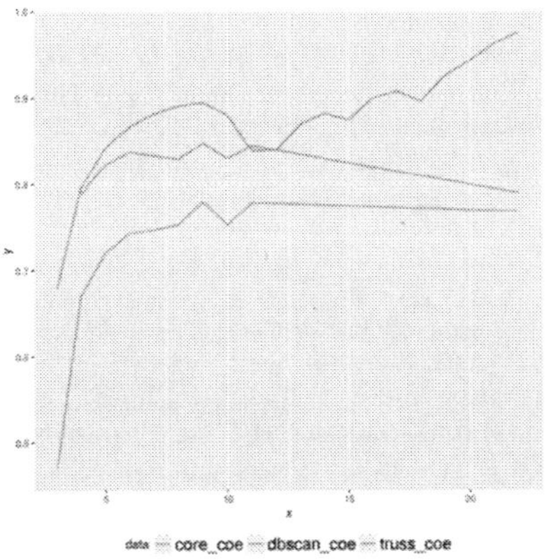


(a) NY BUR

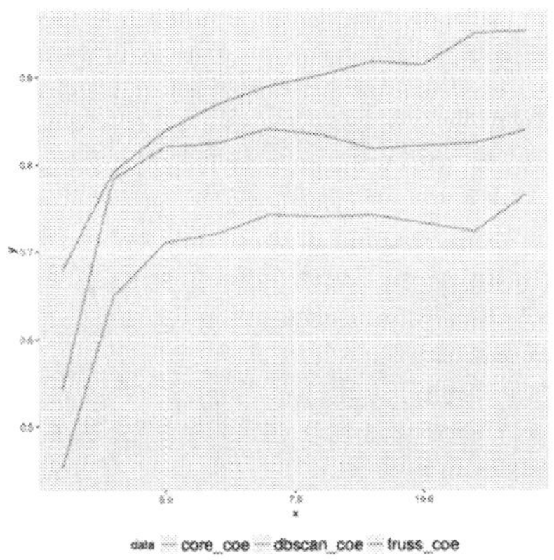


(b) DC BUR

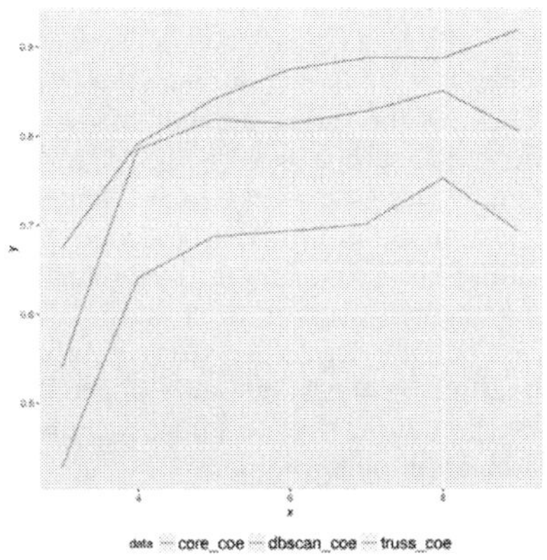


(c) CHI BUR

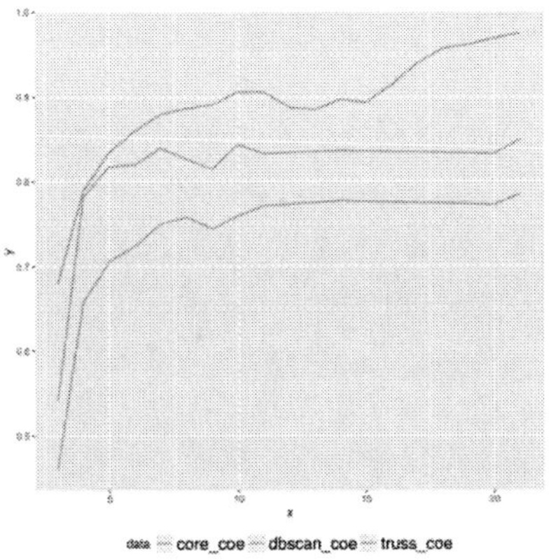


(d) NY ROB

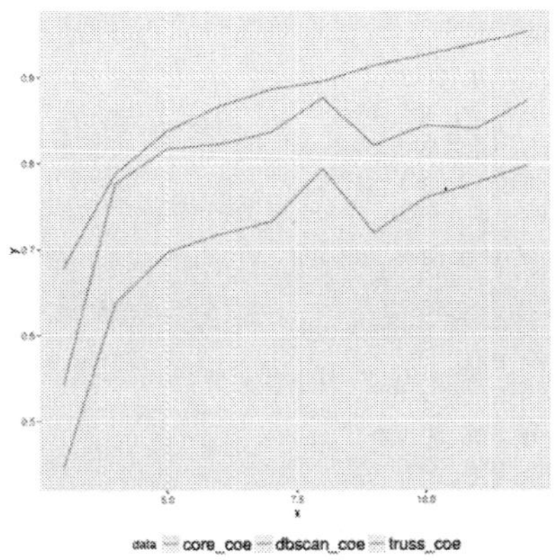


(e) DC ROB

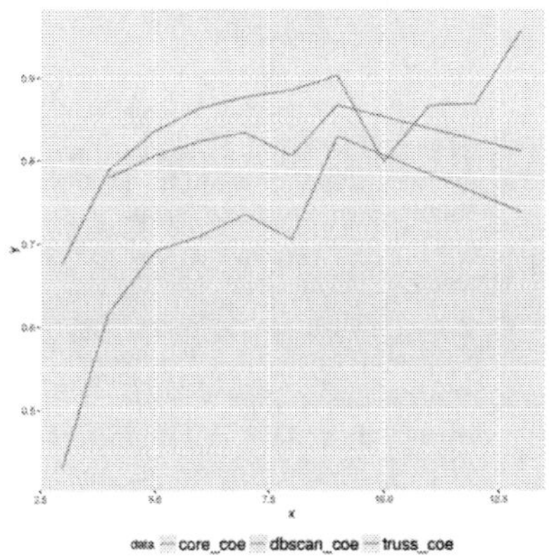


(f) CHI ROB

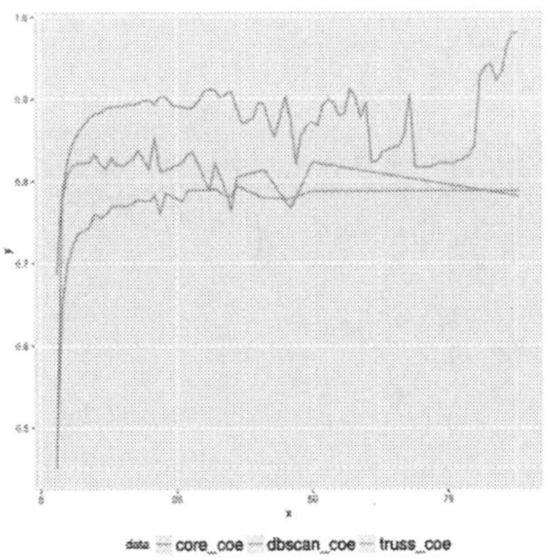


(g) NY TFT

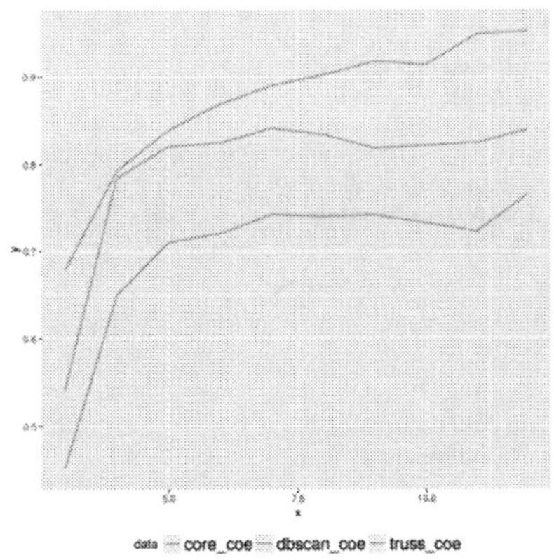


(h) DC TFT

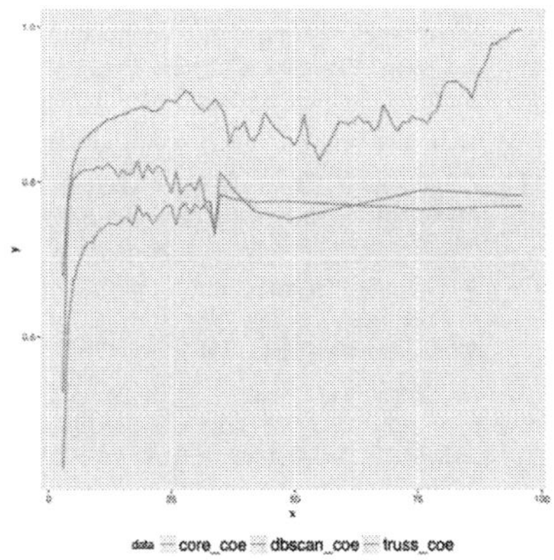


(i) CHI TFT

Figure 8. The clustering coefficient of cohesive subgraphs detected on different data (the x-axis is $k$, and the y-axis is the cluster coefficient value).

event chain numbers such that we will be able to know whether the near-repeat effect also exists in the crime clusters. Meanwhile, we have noticed that there are the sheer amount of duplicated events in the real-world data, while the tightness of relationships between each event pair is not the same. How to handle these conditions in the sense of weighted vertex and edges is a challenging theoretical problem. Besides, when it comes to the larger amount of data, for example, to handle the online crime events, the data size will be much larger than what we process now. Hence, a parallel event chain detection algorithm is also needed. Since the crime events are adding each day, how to dynamically detecting event chains incrementally becomes an issue theoretically and practically. Last but not least, the near-repeat effect is not only existed in crime analysis but also existed in many areas such as transportation, how to utilize our method in other areas is a fascinating open problem.

## REFERENCES


[1] "City of Chicago Data Portal," [retrieved: 02/2020]. [Online]. Available: https:// data.cityofchicago.org/ Public-Safety/Crimes-2001-to-present /ijzp-q8t2/data

[2] "District of Columbia Open Data," [retrieved: 02/2020]. [Online]. Available: http:// opendata.dc.gov/ datasets?q=crime&sort_by=relevance

[3] "Historical New York City Crime Data," [retrieved: 02/2020]. [Online]. Available: http:// www.nyc.gov/html/nypd/html/ analysis_and_planning/ historical_nyc_crime_data.shtml

[4] E. Amazon, "Amazon web services," *Available in: http://aws. amazon. com/es/ec2/(November 2012)*, 2015.

[5] V. Batagelj and M. Zaversnik, "An o (m) algorithm for cores decomposition of networks," *arXiv preprint cs/0310049*, 2003.

[6] D. Birant and A. Kut, "St-dbscan: An algorithm for clustering spatial–temporal data," *Data & Knowledge Engineering*, vol. 60, no. 1, pp. 208–221, 2007.

[7] I. M. Bomze, M. Budinich, P. M. Pardalos, and M. Pelillo, "The maximum clique problem," in *Handbook of combinatorial optimization*. Springer, 1999, pp. 1–74.

[8] M. F. Buchroithner and R. Pfahlbusch, "Geodetic grids in authoritative maps–new findings about the origin of the utm grid," *Cartography and Geographic Information Science*, pp. 1–15, 2016.

[9] J. Cohen, "Trusses: Cohesive subgraphs for social network analysis," *National Security Agency Technical Report*, p. 16, 2008.

[10] T. H. Cormen, *Introduction to algorithms*. MIT press, 2009.

[11] J. Gan and Y. Tao, "Dbscan revisited: Mis-claim, un-fixability, and approximation," in *Proceedings of the 2015 ACM SIGMOD international conference on management of data*, 2015, pp. 519–530.

[12] A. Guttman, "R-trees: A dynamic index structure for spatial searching," in *Proceedings of the 1984 ACM SIGMOD international conference on Management of data*, 1984, pp. 47–57.

[13] X. Huang, H. Cheng, L. Qin, W. Tian, and J. X. Yu, "Querying k-truss community in large and dynamic graphs," in *Proceedings of the 2014 ACM SIGMOD international conference on Management of data*. ACM, 2014, pp. 1311–1322.

[14] X. Huang, W. Lu, and L. V. Lakshmanan, "Truss decomposition of probabilistic graphs: Semantics and algorithms," in *Proceedings of the 2016 International Conference on Management of Data*, 2016, pp. 77–90.

[15] J. Konc and D. Janezic, "An improved branch and bound algorithm for the maximum clique problem," *MATCH Communications in Mathematical and in Computer Chemistry*, vol. 58, no. 3, p. 5, 2007.

[16] J. H. Ratcliffe and G. F. Rengert, "Near-repeat patterns in philadelphia shootings," *Security Journal*, vol. 21, no. 1-2, pp. 58–76, 2008.

[17] Y. Saad and K. SPARS, "A basic tool kit for sparse matrix computations," *RIACS, NA SA Ames Research Center, TR90-20, Moffet Field, CA*, 1990.

[18] N. Satish *et al.*, "Navigating the maze of graph analytics frameworks using massive graph datasets," in *Proceedings of the 2014 ACM SIGMOD international conference on Management of data*, 2014, pp. 979–990.

[19] M. B. Short, M. R. Dorsogna, P. Brantingham, and G. E. Tita, "Measuring and modeling repeat and near-repeat burglary effects," *Journal of Quantitative Criminology*, vol. 25, no. 3, pp. 325–339, 2009.

[20] J. G. Siek, L.-Q. Lee, and A. Lumsdaine, *Boost Graph Library: User Guide and Reference Manual, The*. Pearson Education, 2001.

[21] M. Townsley, "Near repeat burglary chains: describing the physical and network properties of a network of close burglary pairs," in *Crime Hot Spots: behavioral, computation, and mathematical models symposium*, vol. 1, no. 31, 2007, p. 2007.

[22] M. Townsley, R. Homel, and J. Chaseling, "Infectious burglaries. a test of the near repeat hypothesis," *British Journal of Criminology*, vol. 43, no. 3, pp. 615–633, 2003.

[23] J. Wang and J. Cheng, "Truss decomposition in massive networks," *Proceedings of the VLDB Endowment*, vol. 5, no. 9, pp. 812–823, 2012.

[24] D. J. Watts and S. H. Strogatz, "Collective dynamics of small-worldnetworks," *nature*, vol. 393, no. 6684, pp. 440–442, 1998.

[25] W. Wells, L. Wu, and X. Ye, "Patterns of near-repeat gun assaults in houston," *Journal of Research in Crime and Delinquency*, p. 0022427810397946, 2011.

[26] T. White, "Hadoop: The definitive guide," *Oreilly Media Inc Gravenstein Highway North*, vol. 215, no. 11, pp. 1 – 4, 2010.

# EPOS: A FAIR Research Infrastructure

Keith G Jeffery
Keith G Jeffery Consultants
Faringdon, UK
Email: keith.jeffery@keithgjefferyconsultants.co.uk

Daniele Bailo
ERIC
Istituto Nazionale di Geofisica e Vulcanologia
Rome, Italy
Email: daniele.bailo@ingv.it

Kuvvet Atakan
Department of Earth Science
University of Bergen
Bergen, Norway
Email: kuvvet.atakan@uib.no

Matt Harrison
Director Informatics
British Geological Survey
Keyworth, UK
Email: mharr@bgs.ac.uk

*Abstract*—**The European Plate Observing System (EPOS) has been developed over some years and is now in transition to full operational status. It currently offers a portal with access to more than 200 data services, the portals of constituent research communities and prototype access to a service for Trans-National Access (TNA) to equipment and sensors as well as access to information on the organizations and persons involved in EPOS together with research capabilities. From the beginning, EPOS was designed to support the FAIR (Findable, Accessible, Interoperable, Reusable) principles. This paper explains how the EPOS architecture meets the specifications of FAIRness.**

***Keywords- geoscience; data services; metadata; CERIF; catalog; research infrastructures; FAIR.***

## I. INTRODUCTION

The architecture of EPOS was described in [1]. The present work focuses on how FAIR principles are applied in EPOS. The purpose of EPOS is to provide end-users – including researchers, educators, policymakers, industry employees, citizen scientists – with the ability to discover, contextualize and utilize the heterogeneous assets of the various geoscience communities through a homogeneous interface.

### A. Overview

The architecture has been designed to satisfy the following criteria:

1. Minimal interference with existing communities' operations and developments, including Information Technology (IT);
2. Easy-to-use user interface;
3. Access to assets through a metadata catalog: initially services, but progressively also datasets, workflows, software modules, computational facilities, instruments/sensors, all with associated organizational information including experts and service managers;
4. Progressive assistance in composing workflows of services, software and data to deploy on e-Infrastructures to achieve research infrastructure user objectives.

### B. FAIRness

From the beginning, EPOS was designed to be FAIR and EPOS participants were involved in the discussions leading to the FAIR principles [2] and also subsequent work on FAIR metrics within the FAIR Data Maturity Model Working Group of Research Data Alliance (RDA) [3]. The major contributions of this paper are to indicate (a) how FAIRness was achieved from the beginning of EPOS: (b) how our systems development approach maps to the FAIR principles using a 'pyramid' diagram.

### C. Previous Work

EPOS provides an original approach to the provision of homogeneous access over heterogeneous digital assets and providing FAIRness. Previous work on homogenizing heterogeneity has been mainly within a limited domain (where standards for assets and their metadata may be consensual across the whole domain thus reducing heterogeneity) with manual processes and associated costs. Filematch [4] exhibited those problems. NASA has a Common Metadata Repository (CMM). In 2013, NASA developed the Unified Metadata Model (UMM) [5] to and from which, other metadata standards are converted. This follows the superset canonical rich metadata approach already used in EPOS. The Open Geospatial Consortium (OGC) has produced a series of standards. GeoNetwork [6] has established a suite of software based around the OGC ISO19115 metadata standard; however, despite its open nature, this software 'locks in' the developer to a particular way of processing, does not assist in the composition and deployment of workflows and the metadata is insufficiently rich for automated processing. EarthCube [7] is a collection of projects providing designs and tools for geoscience, including interoperability, in USA. The project encountered – by using pairwise brokering – the problem that it required n*(n-1) brokers instead of the n required if a canonical metadata approach is used. Auscope [8] includes AuScope GRID which, by using ISO19115, encounters the problems outlined above. GEOSS [9] uses the 'system of systems' approach, but this requires many bilateral interfaces with the combinatorial problem discussed above.

In essence, all these other approaches provide some degree of FAIRness (Finding, Accessing), but usually require human and manual work to achieve interoperability or reuse.

EPOS, with its superset rich canonical metadata, overcomes the problems concerning homogeneous access over heterogeneous assets and, furthermore, provides increasingly automated FAIRness.

The rest of the paper is organized as follows: Section II describes the architecture; Section III discusses the importance of metadata; Section IV demonstrates that EPOS is FAIR and Section V summarizes conclusions.

## II. ARCHITECTURE

The Information and Communication Technologies (ICT) architecture of EPOS is designed to facilitate the research community and others in discovering and utilizing through the Integrated Core Services (ICS) the assets provided by the Thematic Core Services (TCS) communities. The architecture was described in [1], but is recapitulated briefly for this paper.

### A. *Introduction*

In order to provide end-users with homogeneous access to services and multidisciplinary data collected by monitoring infrastructures and experimental facilities (and to software, processing and visualization tools as well), a complex, scalable and reliable architecture is required. A diagram of the architecture is outlined in Figure 1.

The key aspects are:

1. National Research Infrastructures (NRI) hold the assets and provide metadata to describe them;
2. Thematic Core Services (TCS) that relate to (currently 10) communities, each for a particular domain of geoscience. These communities harmonise progressively semantic aspects of metadata such as terminology in ontologies and also decide which NRI assets should be proposed for availability through EPOS;
3. Integrated Core Services (ICS) that provide the portal, associated metadata catalog and thus provide Findability, Accessibility, Interoperability, and Reusability (FAIR).

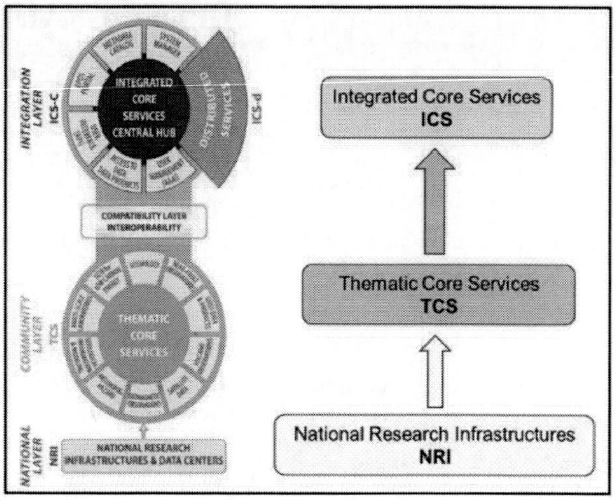


Figure 1. EPOS Architecture.

### B. *ICS*

The EPOS-ICS provides the entry point to the EPOS environment. ICS-C provides the portal and metadata catalog, with associated convertors, to accept metadata from TCS and ingest into the catalog. ICS-D provides distributed computational resources including also processing and visualization services, of which a specialization is Computational Earth Science (CES). ICS-C provides the basis for deployment of workflows, including to ICS-D facilities, that in turn rely on e-Infrastructures such as Cloud Computing or supercomputing. EPOS has also been involved in the VRE4EIC project [10] (and cooperating with EVER-EST [11]) to ensure convergent evolution of the EPOS ICS-C user interface and Application Programming Interfaces (APIs) for programmatic access with the developing Virtual Research Environments (VREs). EPOS participates in the recently approved ENVRIFAIR project [12] that will improve the deployments to the European Open Science Cloud (EOSC) [13] (See Figure 2).

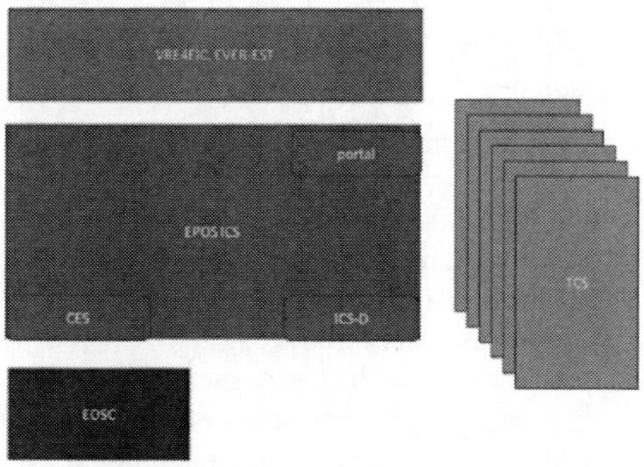


Figure 2. EPOS Positioning.

Workflow for the deployment (which may be a simple file download or a complex set of services including analytics and visualization) will be generated within the ICS-C, by interaction with the users. The workflow will be checked by the end-user before deployment. However, the detailed content/capability of the assets might not be known, e.g., the dataset may not contain the relevant information despite its metadata description, or the software may not execute as the user expects despite the metadata description. The execution of the deployment is monitored and execution information is returned to the end-user. The ICS represents to the end-user the infrastructure, consisting of services that will allow access to multidisciplinary resources provided by the TCS. These will include data and data products as well as synthetic data from simulations, processing, and visualization tools. The key to this view of the geoscience domain is the metadata catalog using the Common European Research Information Format (CERIF) [14].

### *C. ICS-C*

The ICS-C consists of multiple logical areas of functionality, these include the Graphical User Interface (GUI), web-API, metadata catalogue, user management etc. A micro-service architecture has been adopted in the ICS-C, where each (micro) services is atomic and dedicated to a specific class of tasks. The EPOS ICS-C system architecture is outlined in (Figure 3). The Microservices architecture approach envisages small atomic services dedicated to the execution of a specific class of tasks, which have high reliability [15][16]. Docker Containers technology was used. enabling complete isolation of independent software applications running in a shared environment. The communication between microservices is done via messages received and sent on a queueing system, in this case RabbitMQ [17]. As a result, a chain of microservices processes the requests.

The current architecture includes an Authentication, Authorization, Accounting Infrastructure (AAAI). This has been implemented using UNITY [18] and has involved close cooperation with CYFRONET, evolving to the integrated authentication system for research communities. Authorization is more complex, and is being developed incrementally, as it depends on rules agreed with the TCS (within the context of the financial, legal and governance traversal workpackages of EPOS-IP) for each of their assets, and included further metadata elements into the CERIF catalog to control such authorization. The latter has been prepared and awaits validation by the TCS. Related to this, the GUI now provides a user notification pointing to a legal disclaimer for the EPOS system. It should be noted that use of authentication and authorization does not preclude FAIRness, but does allow for protection of assets e.g. to allow a research team time to publish results based on their data before the data is made generally available.

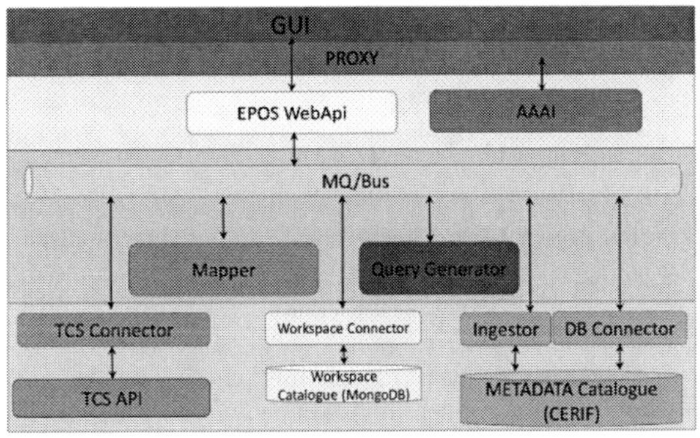


Figure 3. ICS-C Architecture.

The topic of workflow has required pilot projects with TCS experts to clarify the requirements, available technologies and the difficulties of appropriate user interactions. Working in cooperation with the VRE4EIC project we have the basic components for (a) a general workflow manager interface; (b) interfaces to specific workflow managers such as Taverna [19].

### *D. ICS-D*

ICS-D concerns workflow management, since once Found and Accessed, the assets are Interoperable and Reusable, using workflows distributed across e-Infrastructure components by ICS-D. A specification of the metadata elements required for ICS-D has been developed, and is still being refined in the light of experience from the pilots mentioned above. ICS-D will appear to the workflow, or to the end-user, as a service accessed through an API. The deployment requires middleware. Results from the PaaSage project [20] are relevant and the concurrent MELODIC project [21] offers optimization, including that based on dataset placement and latency. Further refinement of requirements and the architectural interfaces continues.

## III. METADATA

The core of the EPOS architecture is the metadata catalog and specifically the superset, rich, canonical metadata format chosen, namely CERIF. This allows EPOS to provide support for cross-domain, interoperable science while achieving the objectives of the FAIR principles.

### *A. Introduction*

The metadata catalogue is the way of representing in a homogeneous way, the heterogeneous assets provided within the EPOS community. The catalog defines what assets are visible to end-users. It provides the required information to facilitate Finding, Accessing, Interoperating and Reusing (FAIR) EPOS assets. In fact, between Finding and Accessing, the use of a rich format like CERIF also allows contextualization: that is the assessment of relevance and quality of the asset for the purpose in hand. Furthermore, the use of linking entities between base entities in CERIF, with role and temporal interval, provides automatically records describing provenance since it is possible to retrieve all link entity records related to a particular base entity, role or time interval in any combination. The catalogue contains: (i) technical specification to enable autonomic ICS access to TCS discovery and access services, (ii) metadata associated with the digital object with a direct link to it, (iii) information about users, resources, software, and services other than data services (e.g., rock mechanics, geochemical analysis, visualization, processing).

The CERIF data model was chosen because it: (1) separates base entities from linking entities, thus providing a fully connected graph structure; (2) using the same syntax, stores the semantics associated with values of attributes, both for base entities and (for role of the relationship) for linking entities, that also store the temporal duration of the validity of the linkage. This provides great power and flexibility. CERIF also (as a superset) can interoperate with widely adopted metadata formats such as Dublin Core (DC) [22], Data Catalogue Vocabulary (DCAT) [23], Comprehensive Knowledge Archive Framework (CKAN) [24], INSPIRE (the EC version of ISO 19115 for geospatial data) [25] and others using convertors developed as required to meet the metadata mappings achieved between each of the above standards and CERIF. Currently 17 different metadata formats in geoscience are convertible with CERIF. The metadata catalogue also manages the semantics, in order to provide the meaning of the attribute values.

To recap, the use of CERIF automatically provides:

(a) The ability for discovery, contextualization, interoperation and (re-)use of assets according to the FAIR principles [2]
(b) A clear separation of base entities (things) from link entities (relationships);
(c) Formal syntax and declared semantics;
(d) A semantic layer, also with the base/link structure allowing crosswalks between semantic terminology spaces;
(e) Conversion to/from other common metadata formats;
(f) Built-in provenance information, because of the timestamped role-based links;
(g) Curation facilities, because of being able to manage versions, replicates and partitions of digital objects using the base/link structure;

These technical properties of CERIF provide that which is required to ensure FAIRness of the system. The catalog is constantly evolving with the addition of new assets (such as services, datasets), but also increasingly rich metadata, as the TCSs improve their metadata collection to enable more autonomic processing.

### B. TCS Metadata

The 'treasure' of EPOS is the assets provided, through the TCS communities from the NRIs. These TCS Data, Data Products, Software and Services (DDSS) are described by metadata. The metadata describing those assets is supplied via the TCS IT experts and is harmonized as much as possible. It is checked for quality, and registered in the granularity database (see below). This relates to governance, including funding for the TCS. It is then converted to CERIF via an intermediate format (see below).

### C. ICS Metadata

The intermediate format is known as the EPOS baseline. It provides a minimum set of common metadata elements required to operate the ICS, taking into consideration the heterogeneity of the assets of the many TCSs involved in EPOS. It has been implemented as an application profile using an extension of the DCAT standard, namely the EPOS-DCAT-AP. The baseline can be extended to accommodate extra metadata elements, where it is deemed that those metadata elements are critical in describing and delivering the data services for any given community. Indeed, this has happened already when the original EPOS-DCAT-AP was found to be inadequate, and a new version with richer metadata was designed and implemented.

The metadata to be obtained from the EPOS TCSs, as described in the baseline document (and any other agreed elements) will be mapped to the EPOS ICS CERIF catalog. The process of converting metadata acquired from the EPOS TCS to CERIF will be done by in consultation with each TCS as to what metadata they have available and harvesting mechanisms

The metadata is ingested from the TCS community NGIs by various mechanisms, depending on local conditions. In general, they expose an API allowing the metadata to be collected. The metadata is transformed from local format to EPOS baseline and thence to CERIF. These APIs, and the corresponding ICS convertors, collectively form the "interoperability layer" in EPOS, which is the link between the TCSs and the ICS.

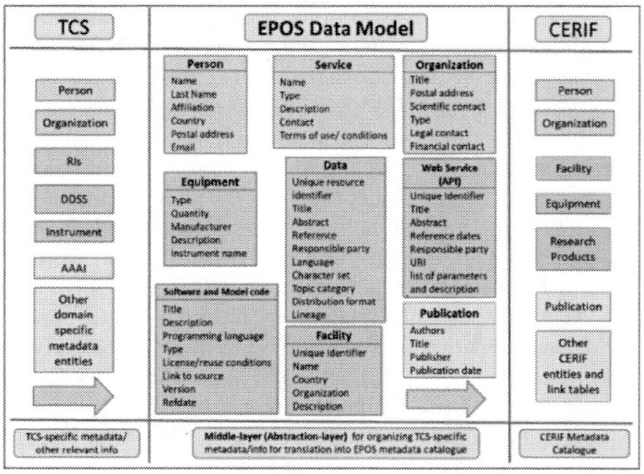


Figure 4. EPOS Metadata Baseline.

The EPOS baseline can thus be considered as an intermediate layer, that facilitates the conversion from the community metadata standards such as ISO19115/19, DCAT, Dublin Core, INSPIRE, etc., describing the DDSS elements and not the index or detailed scientific data (See Figure 4).

*D. DDSS and Granularity Database*

As a part of the Requirements and Use cases Collection (RUC) from the TCSs, a specific list was prepared to include all data, data products, software and services (DDSS). The DDSS master table was originally implemented as Excel spreadsheets. The DDSS Master Table was also used for extracting the level of maturity of the various DDSS elements in each TCS, as well as providing a summary of the status of the TCS preparations for the ICS integration and interoperability. The current version of the DDSS Master Table consists of 363 DDSS elements, where 165 of these already exist and are declared by TCSs to be ready for implementation. The remaining DDSS elements required more time to harmonize the internal standards, prepare an adequate metadata structure and so are available for implementation soon. In total, 21 different harmonization groups (HGs) are established to help organizing the harmonization issues in a structured way. In addition, user feedback groups (UFGs) have been established and work to give constant and structured feedback during the implementation process of the TCS-ICS integration and the development of the ICS.

The rate of change of the DDSS maser table indicated that a different technology should be used. The DDSS master table has been transformed to the granularity database because of the problems of referential and functional integrity using a spreadsheet; relational technology provides appropriate constraints to ensure integrity.

An increasingly detailed RUC collection process is formulated and explained through dedicated guidelines and interview templates. A roadmap for the ICS-TCS interactions for the RUC collection process was prepared for this purpose and distributed to all TCSs.

In this approach, a five-step procedure is applied involving the following:

- Step 1: First round of RUC collection for mapping the TCS assets;
- Step 2: Second round of RUC collection for identifying TCS priorities;
- Step 3: ICS-TCS Integration Workshop for building a common understanding for metadata
- Step 4: Third round of RUC collection for refined descriptions before implementation;
- Step 5: Implementation of RUC to the CERIF metadata;

This procedure has been refined over man months, but is designed to ensure maximum richness, integrity and correctness of the metadata, since it is upon the quality of the metadata that the achievement of FAIRness depends.

Work is now complete in converting the DDSS tables (in Excel) to the granularity database using Postgres. This (a) facilitates finding particular DDSS elements, eliminating duplicates and checking the progress of getting DDSS elements into the metadata format; (b) simplifies harvesting to the metadata catalog.

## IV. DEMONSTRATING THAT EPOS IS FAIR

The mapping of the FAIR principles to aspects of the EPOS architecture, demonstrates that the FAIR principles are supported by the EPOS architecture from metadata to service provision (Figure 5).

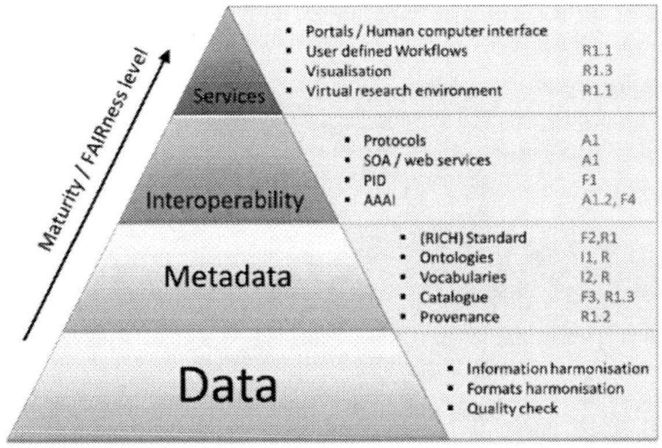


Figure 5. The FAIR Principles and the EPOS Architecture Pyramid [27] [28].

The provision of FAIRness starts with the metadata as explained above. To achieve FAIRness, the metadata must be rich (many attributes), identify uniquely the asset with a Resolvable Universally Unique Persistent IDentifier (RUUPID), have available licensing information, use standard protocols, have an appropriate vocabulary, provide qualified references and provenance. We believe to this should be added demonstrate both referential and functional integrity. It is on the latter two quality measures that many other metadata formats fail.

Findability is achieved by the rich metadata. Query on the rich metadata selects the metadata representing the assets of interest, including the RUUPID of the asset.

Accessibility is achieved by resolving the asset RUUPID and also ensuring the access conditions – in a licence (better a machine-representation of the conditions in the licence) or metadata concerning authorization from the AAAI – are respected.

Interoperability is achieved by the use of convertors between metadata formats, to provide homogeneous access to the assets through standard APIs. If necessary, data formats can also be converted to a canonical form to allow co-analysis or display of heterogeneous datasets.

Reusability is achieved because of the richness of the metadata (many attributes), the provision of licence information allied to the authorization component of AAAI, the utilization of community standard formats and finally the provision of provenance information which comes automatically because of the time-stamped role-based relationships between base entities in CERIF.

## V. CONCLUSION

Currently, 186 digital assets (rising progressively to 221) from the domain communities, supported by 281 webservices, are represented by CERIF metadata in the EPOS ICS-C catalog and made available FAIRly. These services, described by the metadata, can be discovered, accessed, contextualized and (re)utilized individually or composed into workflows and hence become interoperable. A GUI provides the user view onto the catalog, and it also provides a workspace to collect the metadata of the assets selected for use (Figure 6). From the workspace a workflow may be constructed and deployed.

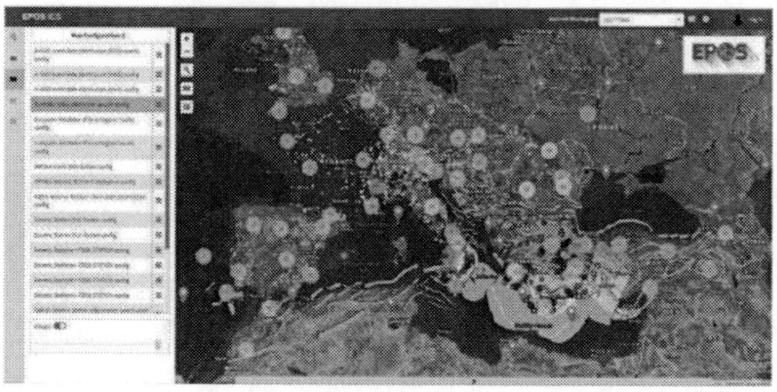

Figure 6. EPOS-ICS Graphical User Interface.

Future plans include:

(a) Harvesting of metadata describing more assets: not only services, but also datasets, software, workflows, equipment;
(b) Improving the GUI to allow workflow deployment with 'fire and forget' technology, or single-step with user checking and adjustment at each step;
(c) Completion of the software to permit trans-national access to laboratory and sensor equipment;
(d) Improved AAAI to give the domain communities finer control over FAIR utilisation of their assets;
(e) The inclusion of virtual laboratory-type interfaces (virtual research environments), allowing users access and connectivity including open-source frameworks such as Jupyter notebooks [26], which are increasingly being used in some scientific communities.

The architecture outlined and demonstrated (in successive prototypes) in EPOS-IP has found favour (not without some criticism of course – leading to agile improvements) from the user community. The criticisms usually concerned: (a) simplifying the complexity of the user interface (achieved by the use of panes); (b) improvements in the quantity (more attributes) and quality of metadata to make Finding, Accessing Interoperating and Reusing easier – this was really a criticism of the TCS supplied metadata more than the ICS; (c) lack of harmonization – again this is the responsibility of harmonization groups across the TCS communities. Furthermore, the prototype system has passed Technological Readiness Assessment procedures within the governance of the EPOS-IP project. Currently the ICS is undergoing pre-production tests. The architecture meets the requirements, it is state of the art and has a further development plan. The FAIR achievements are:

1. EPOS architecture from the beginning was designed for FAIR, with EPOS staff involved in FAIR definition and subsequent indicators work;
2. EPOS is already FAIR-compliant with RUUPIDs, rich metadata (many attributes), formal syntax, declared semantics, referential and functional integrity;
3. The EPOS catalog already interoperates with 17 metadata 'standards' in geoscience and wider;
4. EPOS is open to interoperate with other RIs
(a) directly; (b) via an 'umbrella' VRE; or (c) via EOSC;
5.EPOS started with interoperable services which overcomes many problems with data and is anticipating EOSC.

## ACKNOWLEDGMENT

The authors acknowledge the work of the whole ICT team in EPOS reported here, and the funding of the European Commission H2020 program (Grant agreement 676564) and National Funding Councils that have made this work possible.

## REFERENCES

[1] K. Jeffery, D. Bailo, K. Atakan, and M. Harrison "EPOS: European Plate Observing System" in Proc. Eleventh International Conference on Advanced Geographic Information Systems, Applications, and Services (GEOProcessing 2019), pp. 79-86.
[2] FAIR Principles https://www.force11.org/group/fairgroup/fairprinciples (accessed on 30 January 2020)
[3] RDA Working Group https://www.rd-alliance.org/groups/fair-data-maturity-model-wg (accessed on 30 January 2020)
[4] P. Sutterlin, K. Jeffery, and E. Gill: "Filematch: A Format for the Interchange of Computer-Based Files of Structured Data" Computers and Geosciences 3 1977) pp. 429-468.
[5] UMM: https://earthdata.nasa.gov/about/science-system-description/eosdis-components/common-metadata-repository/unified-metadata-model-umm (accessed on 30 January 2020)
[6] Geonetwork https://geonetwork-opensource.org/ (accessed 30 May 2019)
[7] EarthCube: https://www.earthcube.org/ (accessed on 30 January 2020)
[8] AuScope: http://www.auscope.org.au/ (accessed on 30 January 2020)
[9] GEOSS: https://www.earthobservations.org/geoss.php (accessed on 30 January 2020)9)
[10] VRE4EIC: https://www.vre4eic.eu/ (accessed on 30 January 2020)
[11] EVEREST: https://ever-est.eu/ (accessed on 30 January 2020)
[12] ENVRI-FAIR: http://envri.eu/envri-fair/ (accessed on 30 January 2020)
[13] EOSC: https://ec.europa.eu/research/openscience/index.cfm?pg=open-science-cloud (accessed on 30 January 2020)

[14] CERIF: https://www.eurocris.org/cerif/main-features-cerif (accessed on 30 January 2020)
[15] Newman, Sam. "Building Microservices", O'Reilly Media, Inc., 2015
[16] International Journal of Open Information Technologies ISSN: 2307- 8162
[17] RabbitMQ: https://www.rabbitmq.com/ (accessed on 30 January 2020)
[18] UNITY: http://www.unity-idm.eu (accessed on 30 January 2020)
[19] Taverna: https://taverna.incubator.apache.org/ (accessed on 30 January 2020) [20] PaaSage: https://paasage.ercim.eu/ (accessed on 30 January 2020)
[21] MELODIC: melodic.cloud/ (accessed on 30 January 2020)
[22] DC: http://dublincore.org/documents/dces/ (accessed on 30 January 2020)
[23] DCAT: https://www.w3.org/TR/vocab-dcat/ (accessed on 30 January 2020)
[24] CKAN: https://ckan.org/ (accessed on 30 January 2020)
[25] INSPIRE: https://inspire.ec.europa.eu/ (accessed on 30 January 2020)
[26] Jupyter: https://jupyter.org/ (accessed on 30 January 2020)
[27] D. Bailo, (2019, July 10). "Four-stages FAIR Roadmap - FAIR "Pyramid"". Zenodo. http://doi.org/10.5281/zenodo.3299353 (accessed on 30 January 2020)
[28] D. Bailo, R. Paciello, M. Sbarra, R. Rabissoni, V. Vinciarelli and M. Cocco "Perspectives on the Implementation of FAIR Principles in Solid Earth Research Infrastructures" , Frontiers in Earth Science, vol 8, 2020, p.3, DOI10.3389/feart.2020.00003 https://www.frontiersin.org/article/10.3389/feart.2020.00003 (accessed on 30 January 2020)

# From Knowledge and Meaning Towards Knowledge Pattern Matching: Creating, Processing, and Developing Knowledge Objects, Targeting Geoscientific Context and Georeferencing

Claus-Peter Rückemann
Westfälische Wilhelms-Universität Münster (WWU), Germany;
Knowledge in Motion, DIMF, Germany;
Leibniz Universität Hannover, Germany
Email: ruckema@uni-muenster.de

*Abstract*—**This paper presents the results of the long-term research on advanced knowledge based mining enabled by conceptual knowledge frameworks. The paper presents the methodological base of a new algorithm framework of conceptual knowledge pattern matching, allowing the consideration of complementary and descriptive knowledge of meaning and intrinsic object properties. The research is illustrated by practical implementations of knowledge pattern matching, including processing and developing multi-disciplinary and multi-lingual knowledge object entities and resources. Examples in this specialised research concentrate on geoscientific context and georeferencing. The goal of this fundamental research is to create methods of knowledge pattern matching usable with many resources and data collections. The implemented practical approaches are for the first time publicly available with this paper.**

***Keywords–Conceptual Knowledge Pattern Matching Methodology; Superordinate Knowledge Methodology; Advanced Data-centric Computing; UDC; Geoscientific and Geospatial Scenarios.***

## I. INTRODUCTION

Knowledge Mining is supported by a number of common methods and algorithms, e.g., string pattern matching algorithms, associative, comparative, and phonetic algorithms. All these achievements deal with distinct extrinsic properties of respective entities in very limited ways.

The motivation for this research was the lack of suitable facilities for an advanced matching of 'meaning' when creating mining solutions in context of complex multi-disciplinary and multi-lingual Knowledge Resources. Knowledge, meaning, and patterns form relations, which may require some introduction.

The concept of meaning differs from the concept of signification. Semantic and syntactic structures do not suffice to determine the discursive meaning of an expression [1]. Discourse means a way of speaking. On the one hand, grammatically correct phrases may lack discursive meaning. On the other hand, grammatically incorrect sentences may be discursively meaningful. Knowledge and meaning are closely tied with intrinsic and extrinsic properties. Therefore, understanding of intrinsic and extrinsic properties of entities is significant for any context. This is nevertheless true for any case of natural language, esp. considering langage, langue, and parole [2].

Creating practical approaches requires algorithms. An algorithm is a process or set of rules to be followed in problem-solving operations. In general, algorithms cannot, by their fundamental nature, handle intrinsic and extrinsic properties to the same quality and extent. For example, an intrinsic property of a word object is the meaning in mind, the 'lemma'. An extrinsic property of a word object can be a written word. Extrinsic properties do not reflect meaning and insight as their representations do not generally allow reasonable results. Best practice provides us with solid, complementary knowledge concepts and methodologies allowing to create advanced methods.

Data do not have or carry meaning. Therefore, understanding meaning is of major significance in information science when dealing with improving formalisation processes and creating 'logos based' analogies along with cognitive processes. Commonly, cognition (cognitio, from Latin cognoscere, "get to know") is the mental action or process of acquiring knowledge and understanding through thought, experience, and the senses (Source: Oxford dictionary). Analogy (from Greek analogia, ἀναλογία, "proportion") is a cognitive process of transferring information or 'meaning' from a particular subject, the analogue or source, to another, the target.

Nevertheless, aspects of meaning can be described using knowledge complements, e.g., considering factual, conceptual, procedural, and metacognitive knowledge [3]. Especially, conceptual knowledge can relate to any of factual, conceptual, and procedural knowledge. To a comparable extent, metacognitive knowledge can relate to any of factual, conceptual, and procedural knowledge. A practical approach for knowledge pattern matching will be presented in the following sections.

The rest of this paper is organised as follows. Section II introduces the previous work and components. Sections III and IV present methodology, method, and implementation. Sections V and VI present the matching process and resulting tables. Section VII summarises conclusions and future work.

## II. PREVIOUS WORK, COMPONENTS, AND RESOURCES

The fundaments of terminology and understanding knowledge are layed out by Aristotle being an essential part of 'Ethics' [4]. Information sciences can very much benefit from Aristotle's fundaments and a knowledge-centric approach [3] but for building holistic and sustainable solutions, supporting a modern definition of knowledge [5], they need to go beyond the available technology-based approaches and hypothesis [6] as analysed in Platon's Phaidon.

Making a distinction and creating interfaces between methods and the implementation applications, the results of this research are illustrated here along with the practical example of the Knowledge Mapping methodology [7] enabling the creation of new object and entity context environments, e.g., implementing methods for knowledge mining context.

The means to achieve such recommendations even for complex scenarios is to use the principles of Superordinate Knowledge, integrating arbitrary knowledge. The core assembly elements of Superordinate Knowledge are methodology, implementation, and realisation [8]. Separation and integration of assemblies have proven beneficial for building solutions with different disciplines and different levels of expertise. Comprehensive focussed subsets of conceptual knowledge can also provide excellent modular and standardised complements for information systems component implementations, e.g., for environmental information management and computation [9].

For the implementation of case studies, the modules are built by support of a number of major components and resources, which can be used for a wide range of applications, e.g., creation of resources and extraction of entities. The Universal Decimal Classification (UDC) [10] is the world's foremost document indexing language in the form of a multi-lingual classification scheme covering all fields of knowledge and constitutes a sophisticated indexing and retrieval tool. The UDC is designed for subject description and indexing of content of information resources irrespective of the carrier, form, format, and language. UDC is an analytico-synthetic and facetted classification. It uses a knowledge presentation based on disciplines, with synthetic features. UDC schedules are organised as a coherent system of knowledge with associative relationships and references between concepts and related fields. The UDC allows an efficient and effective processing of knowledge data and provides facilities to obtain a universal and systematical view on classified objects. UDC-based references in this publication are taken from the multi-lingual UDC summary [10] released by the UDC Consortium under a Creative Commons license [11]. Facets can be created with any auxiliary tables, e.g., auxiliaries of place and space, time, language, and form as well as general characteristics, e.g., properties, materials, relations, processes, and operations, persons and personal characteristics. Module examples are employing Perl Compatible Regular Expressions (PCRE) [12] syntax for specifying common string patterns and Perl [13] for component wrapping purposes with this case study.

## III. Methodology and Implementation

The implementation strictly follows the fundamental methodological algorithm base.

### A. Methodological algorithm base

The Conceptual Knowledge Pattern Matching (CKPM) methodology targets providing and accessing knowledge object patterns. This methodological algorithm framework is based on the Superordinate Knowledge Methodology, which allows systematical use and thorough processing by the steps:

1) Selecting knowledge objects.
2) Accessing knowledge object patterns.
3) Thorough processing of object entities and references.
4) Object entity analysis, knowledge complements' based.
5) Result formation.

The respective accessing includes the organisation and structures used with the objects and entities. Object patterns need to be accessible to an extent and quality, which allows a sufficient processing for the respective scenario. The requirements for specific scenarios will therefore be individual. The processing includes making use of the characteristics and features of the respective implementations of the knowledge based frameworks providing a conceptual base for a certain method. The conceptual knowledge complements referred from knowledge objects can have their origins from manual as well as from automated processes. For the implementation and realisation, the framework providing the base conceptual knowledge reference patterns is the UDC. The results in this publication use the UDC Summary Linked Data (Main Tables, [14]). Creating facets and patterns can also make use of the common auxiliary signs being part of the UDC framework [15]. The following advanced employment of conceptual knowledge (UDC) is far beyond common application of universal classification.

### B. Implemented method

An implementation of a CKPM based method requires accessible objects and a suitable conceptual framework base for processing and automation. The methodic implementation illustrated here enables to employ an UDC framework appropriate for systematical use, implemented by the steps:

1) Knowledge Resources' objects.
2) Accessing formalised conceptual knowledge object pattern description based on UDC, e.g., including geoscientific context and georeferencing.
3) Processing procedure via pipelines, employing UDC knowledge and forks.
4) Entity analysis, based on UDC framework references.
5) Result formation on base of Knowledge Resources' objects, retaining conceptual knowledge.

In this case, meaning is described by conceptual patterns, which can be searched and analysed. Processing algorithms can follow the given organisation, e.g., the decimal organisation of the UDC, following available forks as will be illustrated for the matching process in the following sections. Processing and analysis includes primary, decimal conceptual knowledge and associated multi-dimensional knowledge in context of the object entities. The method allows advanced data-centric computing procedures. In practice, the facility for consistently describing knowledge is a valuable quality, esp., conceptual knowledge, e.g., using the UDC and its editions.

### C. Implemented conceptual knowledge framework and target

Targeting practical use for advanced geoscientific information and expert systems, conceptual geoscientific and geographic mapping and referencing are required. Geographic conceptual knowledge pattern entities are created based on UDC code references [16] of geography, biography, history. Table I shows an implementation excerpt.

Table I. Conceptual Knowledge Pattern Matching: Impl. UDC references of geography, biography, history (excerpt).

| *Code / Sign Ref.* | *Verbal Description (EN)* |
|---|---|
| UDC:902 | Archaeology |
| UDC:903 | Prehistory. Prehistoric remains, artefacts, antiquities |
| UDC:904 | Cultural remains of historical times |
| UDC:908 | Area studies. Study of a locality |
| UDC:91 | Geography. Exploration of the Earth and of individual countries. Travel. Regional geography |
| UDC:912 | Nonliterary, nontextual representations of a region |
| UDC:92 | Biographical studies. Genealogy. Heraldry. Flags |
| UDC:93/94 | History |
| UDC:94 | General history |

A geoscientific / archaeology example from the case studies and implementations for geoscientific information systems and application components is used for illustration in the next sections. The example will show a tiny subset of the comprehensive, universal conceptual knowledge used.

The above conceptual knowledge contains all the references for geographic context, which includes the conceptual knowledge regarding geographic data, e.g., geoinformation and geodescriptive knowledge. The relevant conceptual knowledge required for geoscientific context is provided by references from natural sciences' context. Any of the conceptual knowledge can be used in any stage of a CKPM process, e.g., in start, intermediate, and target specifications.

Natural sciences related conceptual knowledge pattern entities are created based on UDC code references [17] of mathematics and natural sciences. An excerpt of the implementation is shown in Table II.

TABLE II. CONCEPTUAL KNOWLEDGE PATTERN MATCHING: IMPL. UDC REFERENCES OF MATHEMATICS AND NATURAL SCIENCES (EXCERPT).

| *Code / Sign Ref.* | *Verbal Description (EN)* |
|---|---|
| UDC:51 | Mathematics |
| UDC:52 | Astronomy. Astrophysics. Space research. Geodesy |
| UDC:53 | Physics |
| UDC:54 | Chemistry. Crystallography. Mineralogy |
| UDC:55 | Earth Sciences. Geological sciences |
| UDC:550.3 | Geophysics |
| UDC:551 | General geology. Meteorology. Climatology. |
| UDC:551.21 | Vulcanicity. Vulcanism. Volcanoes. Eruptive phenomena. Eruptions |
| UDC:551.7 | Historical geology. Stratigraphy. Palaeogeography |
| UDC:551.8 | Palaeogeography |
| UDC:551.24 | Geotectonics |
| UDC:56 | Palaeontology |
| UDC:57 | Biological sciences in general |
| UDC:58 | Botany |
| UDC:59 | Zoology |

Time related conceptual knowledge pattern entities are created based on UDC code references [18], especially the auxiliaries of time). Table III shows an implementation excerpt.

TABLE III. CONCEPTUAL KNOWLEDGE PATTERN MATCHING: IMPLEMENTED UDC REFERENCES, AUXILIARIES OF TIME (EXCERPT).

| *Code / Sign Ref.* | *Verbal Description (EN)* |
|---|---|
| UDC:"0" | First millennium CE |
| UDC:"1" | Second millennium CE |
| UDC:"2" | Third millennium CE |
| UDC:"3/7" | Time divisions other than dates in Christian (Gregorian) reckoning |
| UDC:"3" | Conventional time divisions and subdivisions: numbered, named, etc. |
| UDC:"4" | Duration. Time-span. Period. Term. Ages and age-groups |
| UDC:"5" | Periodicity. Frequency. Recurrence at specified intervals. |
| UDC:"6" | Geological, archaeological and cultural time divisions |
| UDC:"61/62" | Geological time division |
| UDC:"63" | Archaeological, prehistoric, protohistoric periods and ages |
| UDC:"7" | Phenomena in time. Phenomenology of time |

Spatial conceptual knowledge pattern entities are created based on UDC code references [19], especially the auxiliaries of spatial features and place (UDC (1/9)), (Table IV).

TABLE IV. CONCEPTUAL KNOWLEDGE PATTERN MATCHING: IMPL. UDC REFERENCES, AUXILIARIES OF SPATIAL FEATURES AND PLACE (EXCERPT).

| *Code / Sign Ref.* | *Verbal Description (EN)* |
|---|---|
| UDC:(1) | Place and space in general. Localization. Orientation |
| UDC:(2) | Physiographic designation |
| UDC:(3) | Places of the ancient and mediaeval world |
| UDC:(31) | Ancient China and Japan |
| UDC:(32) | Ancient Egypt |
| UDC:(33) | Ancient Roman Province of Judaea. The Holy Land. Region of the Israelites |
| UDC:(34) | Ancient India |
| UDC:(35) | Medo-Persia |
| UDC:(36) | Regions of the so-called barbarians |
| UDC:(37) | Italia. Ancient Rome and Italy |
| UDC:(38) | Ancient Greece |
| UDC:(399) | Other regions. Ancient geographical divisions other than those of classical antiquity |
| UDC:(4) | Europe |
| UDC:(5) | Asia |
| UDC:(6) | Africa |
| UDC:(7) | North and Central America |
| UDC:(8) | South America |
| UDC:(9) | States and regions of the South Pacific and Australia. Arctic. Antarctic |

Knowledge Resources' objects carry respective conceptual UDC facets and references, including georeferences.

## IV. BASIC PRINCIPLE PROCESSING IMPLEMENTATION

Regarding an implementation ('lxgrep'), a basic routine preparing object entity input into a common structure is illustrated in Figure 1.

```
if (/^(\S)(.*)/./^                    /||/^$/||/^ *$/) {
  s/^(\S.*)\n/\@ENTRY\@$1@@/;
  s/^(.*)\n/\1@@/;
  s/\@ENTRY\@/\n/;
  open(TMPFILE,">>$tempfile"); print TMPFILE "$_"; close(
   TMPFILE);
}
```

Figure 1. Basic routine preparing input entries (excerpt).

An associated elementary system call implementing a basic regular search is shown in Figure 2.

```
system("egrep␣-h␣$temppat␣$ARGV[0].tmp␣>␣$ARGV[0].grep.
 tmp");
system("mv␣$ARGV[0].grep.tmp␣$ARGV[0].tmp");
```

Figure 2. Elementary system call for a basic regular search (excerpt).

An element for a simple system sort based function used with the above search is shown in Figure 3.

```
print "\tsorting␣entries...\n";
system("sort␣-f␣-k␣1,14␣<$tempfile>$tempfile.out");
unlink $tempfile;
```

Figure 3. Element of simple system call sort function (excerpt).

A simple backformatting routine is given in Figure 4.

```
print "\tbackformatting␣entries...\n";
system("perl␣-e␣'while␣(<>){s/@@/\n/g;chop;print␣$_}'␣<
 $ARGV[0].tmp>$ARGV[0].sort");
unlink "$ARGV[0].tmp";
```

Figure 4. Simple backformatting routine (excerpt).

For further structural, technical details, and pipelining please see the references for the case studies given in the text.

## V. New Matching Process and Processing

The new framework of the matching process and processing includes following the conceptual knowledge forks. Here, the primary, decimal reference forks of the UDC are used for implementation, which provide the red line forks within universal knowledge. Matching is beyond non-conceptual knowledge, e.g., natural language processing and string pattern matching. Especially, country and border concepts cannot be used for specification, e.g., ancient and modern border lines fail to be useful. The process enables places in ancient Greece and Rome, from archaeological and prehistoric times associated with places in the ancient and modern world to be described, e.g., references of the type `UDC:..."63"(37)` and `UDC:..."63"(38)`. Trigger question can be 'Can archaeological artefacts' objects of a certain context be associated with earth science objects?'. A symbolic writing specifying a conceptual expression is shown in Figure 5.

```
STRT:[UDC:.*?90]
CTXT:[[UDC:.*?\(.*?38.*?\)]|[UDC:.*?\"6.*?\"]].*[[UDC
 :.*?\"6.*?\"]|[UDC:.*?\(.*?38.*?\)]]
SRCH:[[UDC:.*?55]|[UDC:.*?912]]
```

Figure 5. Example for symbolic writing of pattern expression (excerpt).

A systematic concept of conceptual knowledge implementation allows advanced features, e.g., pattern range variations, pattern permutations. A basic serial pipeline implementation example test for knowledge objects in `<input>` is shown in Figure 6.

```
cat <input> | lxgrep "'%%IML:.*?UDC:.*?\(38.*?\)'" |
 lxgrep "'%%IML:.*?UDC:.*?\"6.*?\"'" <outputctxt>
cat <outputctxt> | lxgrep "'%%IML:.*?UDC:.*?90'"
cat <outputctxt> | lxgrep "'%%IML:.*?UDC:.*?55'" | lxgrep
  "'%%IML:.*?UDC:.*?912'" | lxgrep LATLON:
```

Figure 6. Example for serial pipeline implementation (excerpt).

The pipeline includes objects containing and referring to latitude / longitude objects. The trackable spatial/place related fork process within the conceptual pattern entity group is illustrated in Figure 7.

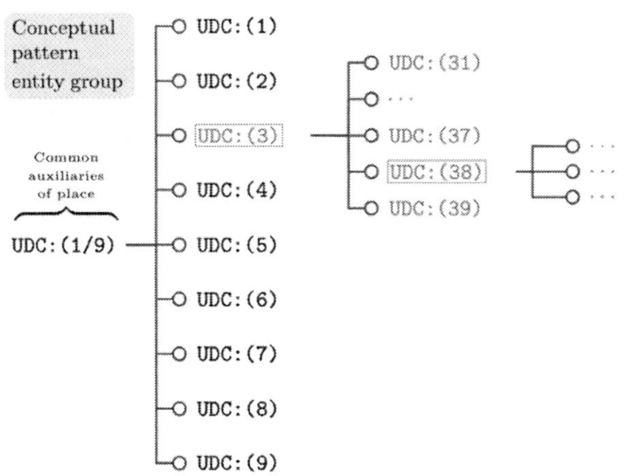


Figure 7. Matching process: Primary, decimal (UDC) conceptual knowledge forks, auxiliaries of spatial features and place (excerpt).

The processing successfully follows the "Ancient Greece" fork. Figure 8 illustrates the fork process within the conceptual pattern entity group for the related conceptual knowledge regarding time.

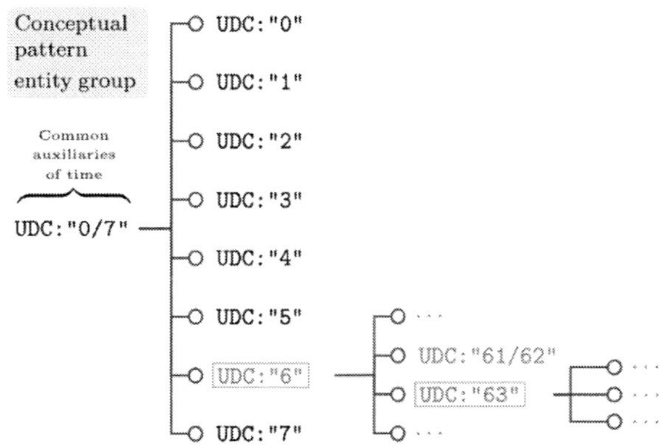


Figure 8. Matching process: Primary, decimal (UDC) conceptual knowledge forks, auxiliaries of time (excerpt).

The processing successfully follows the "geographical/historical" and "natural sciences" fork. The main tables of the conceptual knowledge are managed in the same way within the respective conceptual pattern entity groups (Figure 9).

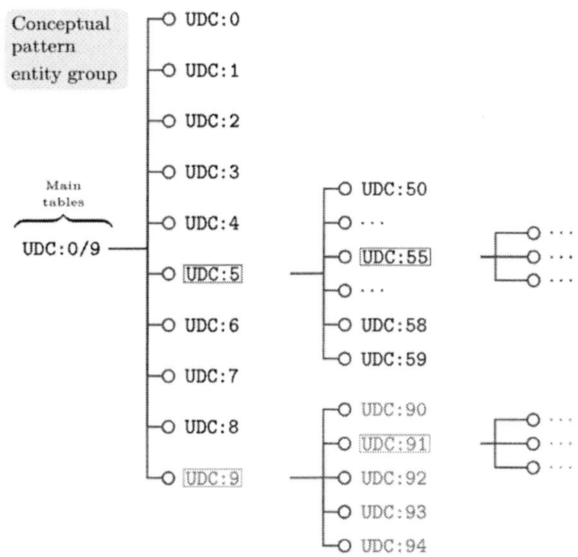


Figure 9. Matching process: Primary, decimal (UDC) conceptual knowledge forks, main tables, including earth sciences and geography (excerpt).

The processing successfully follows the "Earth sciences" and "Geography" forks. These procedures referencing to a formalised [20], practical framework of conceptual knowledge embrace all the relevant universal knowledge, e.g., including natural sciences and geosciences, archaeology, philosophy, and history. The results of removing in the domain of knowledge and removing in the domain of mathematics are not the same. In principle, abstraction means removing [21]. In the mathematical domain, removing is mostly formalised by subtraction [22]. In general, any universal conceptual knowledge framework can be used, which enables a systematical processing and which is universal and consistent.

## VI. Resulting Match Tables

Following the above archaeology-geosciences case of matching process and processing, the resulting match tables contain the references to conceptional and associated multi-dimensional knowledge in context of the object entities. The resulting start match table of object entities (Table V) contains entities and references on details of mythological and archaeological context.

TABLE V. RESULTING CONCEPTUAL KNOWLEDGE PATTERN MATCHING INTERMEDIATE START ('UDC:90') MATCH TABLE (EXCERPT).

| *Object Entity* | *Reference Data (excerpt)* |
|---|---|
| Poseidon | DESC MYTH SYN LOC UDC ... CITE:[23], [24], [25], [26] |
| Polybotes/-is | DESC MYTH SYN LOC UDC ... CITE:[23], [25] |
| Polyvotes/-is | DESC MYTH SYN LOC UDC ... CITE:[23], [25] (transcr.) |

These entities contain descriptions, including transcriptions, transliterations, translations, mythology references, synonyms, location references, UDC references, and citation sources. The citations refer to respective associations of the figured programme with Poseidon and the giant Polybotes / Polybotis / Polyvotes / Polyvotis and further references to the details of mythological context of realia objects, respectively to Parthenon metopes (Acropolis, Athens). The result match table of object entities (Table VI) contains entities and references on details of natural sciences context and georeferences.

TABLE VI. RESULTING CONCEPTUAL KNOWLEDGE PATTERN MATCHING INTERMEDIATE RESULT ('UDC:55') MATCH TABLE (EXCERPT).

| *Object Entity* | *Reference Data (excerpt)* |
|---|---|
| Kos | DESC VOLC VNUM GRC LATLON UDC ... |
| Methana | DESC VOLC VNUM GRC LATLON UDC ... |
| Milos | DESC VOLC VNUM GRC LATLON UDC ... |
| Nisyros | DESC VOLC VNUM GRC LATLON UDC ... |
| Santorini | DESC VOLC VNUM GRC LATLON UDC ... |
| Yali | DESC VOLC VNUM GRC LATLON UDC ... |

The entities in the respective match tables contain descriptions, volcanological references, volcano numbers, country references, latitude and longitude location references, UDC references, and further references. A resulting object is shown in Figure 10. Its media object entities refer to archaeology associated with Poseidon and Polyvotis.

```
Nisyros                 [Volcanology, Geology, Archaeology]:
                        Volcano, Type: Strato volcano, Island,
                        Country: Greece, Subregion Name: Dodecanese Islands,
                        Status: Historical, Summit Elevation: 698\UD{m}. ...
                        Craters: ..., VNUM: 0102-05=, ...
                        %%IML: UDC:[911.2+55],[930.85],[902]"63"(4+38+23+24)=14
                        %%IML: UDC:[912] ...
                        %%IML: media:...{UDC:[911.2+55],"63"(4+38+23)}...jpg
                        Stefanos Crater, Nisyros, Greece.
                        LATLON: 36.578345,27.1680696
                        %%IML: GoogleMapsLocation: https://www.google.com/...@36
                         .578345,27.1680696,337m/...
                        Little Polyvotis Crater, Nisyros, Greece.
                        LATLON: 36.5834105,27.1660736 ...
```

Figure 10. Result object entity from Knowledge Resources: Nisyros object, Greece, containing media object entities and georeferences (excerpt).

As requested, the object contains / refers to latitude / longitude and conceptual knowledge, together with factual knowledge and media references. Figure 11 shows media object entities based on the conceptual knowledge pattern matching process, an object entity at process start (Figure 11(a)), from archaeological artefacts, and a resulting reference object (Figure 11(b)), from natural objects. The media object entities and their context represent the result of the requested knowledge pattern matching, including respective georeferencing properties.

## VII. Conclusion

This research achieved the goal to create a new method of knowledge pattern matching based on the CKPM methodology. The knowledge based mining implementation employed the UDC references in order to provide the required conceptual framework. The UDC references proved to provide an excellent core component, for universal, multi-disciplinary, and multi-lingual knowledge.

In this new context, UDC showed to have a perfect organisational structure of conceptual knowledge for practical, systematical use as well as for an efficient and flexible processing support, following respective knowledge forks for references while creating and keeping developing resources and conceptional knowledge consistent supported by its editions.

In daily practice, the new method provides excellent and sustainable conceptual documentation and enables to create associations and links between knowledge object entities, which cannot result otherwise. Further, configuring knowledge ranges can be achieved in many ways, e.g., by limiting resources, configuring the pattern depths and widths, ranking and selection.

Future research on theory and practice will continue developing suitable knowledge resources and knowledge patterns.

## Acknowledgements

We are grateful to the "Knowledge in Motion" (KiM) long-term project, Unabhängiges Deutsches Institut für Multidisziplinäre Forschung (DIMF), for partially funding this research, implementation, case studies, and publication under grants D2018F6P04938 and D2018F3P04932 and to its senior scientific members and members of the permanent commission of the science council, especially to Dr. Friedrich Hülsmann, Gottfried Wilhelm Leibniz Bibliothek (GWLB) Hannover, to Dipl.-Biol. Birgit Gersbeck-Schierholz, Leibniz Universität Hannover, to Dipl.-Ing. Martin Hofmeister, Hannover, and to Olaf Lau, Hannover, Germany, for fruitful discussion, inspiration, practical multi-disciplinary case studies, and the analysis of advanced concepts. We are grateful to Dipl.-Ing. Hans-Günther Müller, HPE, Germany, for his excellent contributions and assistance providing practical private cloud and storage solutions. We are grateful to the members of the Eastern Mediterranean research and studies campaign 2018–2020, DIMF, and all national and international partners in the Geo Exploration and Information cooperations for their constructive and trans-disciplinary support. We are grateful to the Science and High Performance Supercomputing Centre (SHPSC) for long-term support. / DIMF-PIID-DF98_007.

## References

[1] M. Foucault, The Archaeology of Knowledge. Routledge Classics, 2002, ISBN: 978-0-415-28752-4, Translated by A. M. Sheridan Smith.

[2] F. de Saussure, Cours de linguistique générale, 1916, (title in English: Course in General Linguistics), Charles Bally and Albert Sechehaye (eds.).

(a) Metope, New Acropolis Museum, Athens, (CPR, DIMF, 2019).

(b) Volcano crater, island of Nisyros, Dodecanese Islands, Greece, (CPR, DIMF, 2019).

Figure 11. Result based on the conceptual knowledge pattern matching process, via intermediate match table (Table VI): (a) an artefact, metope (EAST VI), Parthenon, (Archaeology Digital Object Archive, 2019), and a resulting georeferenced object, (b) a natural object (Geosciences Digital Object Archive, 2019).


[3] L. W. Anderson and D. R. Krathwohl, Eds., A Taxonomy for Learning, Teaching, and Assessing: A Revision of Bloom's Taxonomy of Educational Objectives. Allyn & Bacon, Boston, MA (Pearson Education Group), USA, 2001, ISBN: 978-0801319037.

[4] Aristotle, The Ethics of Aristotle, 2005, Project Gutenberg, eBook, EBook-No.: 8438, Rel. Date: Jul., 2005, Digit. Vers. of the Orig. Publ., Produced by Ted Garvin, David Widger, and the DP Team, Edition 10, URL: http://www.gutenberg.org/ebooks/8438 [accessed: 2020-01-12].

[5] C.-P. Rückemann, F. Hülsmann, B. Gersbeck-Schierholz, P. Skurowski, and M. Staniszewski, Knowledge and Computing. Post-Summit Results, Delegates' Summit: Best Practice and Definitions of Knowledge and Computing, Sept. 23, 2015, The Fifth Symp. on Adv. Comp. and Inf. in Natural and Applied Sciences (SACINAS), The 13th Int. Conf. of Num. Analysis and Appl. Math. (ICNAAM), Sept. 23–29, 2015, Rhodes, Greece, 2015, DOI: 10.15488/3409.

[6] Plato, Phaedo, 2008, (Written 360 B.C.E.), Translated by Benjamin Jowett, Provided by The Internet Classics Archive, URL: http://classics.mit.edu/Plato/phaedo.html [accessed: 2020-01-12].

[7] C.-P. Rückemann, "Methodology of Knowledge Mapping for Arbitrary Objects and Entities: Knowledge Mining and Spatial Representations – Objects in Multi-dimensional Context," in Proceedings of The Tenth International Conference on Advanced Geographic Information Systems, Applications, and Services (GEOProcessing 2018), March 25–29, 2018, Rome, Italy. XPS Press, Wilmington, Delaware, USA, 2018, pp. 40–45, ISSN: 2308-393X, ISBN: 978-1-61208-617-0, URL: http://www.thinkmind.org/index.php?view=article&articleid=geoprocessing_2018_3_20_30078 [accessed: 2020-01-12].

[8] C.-P. Rückemann, "Superordinate Knowledge Based Comprehensive Subset of Conceptual Knowledge for Practical Mathematical-Computational Scenarios," in The Ninth Symp. on Adv. Comp. and Inf. in Natural and Applied Sciences (SACINAS), Proceedings of The 17th Int. Conf. of Num. Analysis and Appl. Math. (ICNAAM), Sept. 23–28, 2019, Rhodes, Greece, American Inst. of Physics Conf. Proc. AIP Press, Melville, New York, USA, 2020, ISSN: 0094-243X, (to appear).

[9] C.-P. Rückemann, Sustainable Knowledge and Resources Management for Environmental Information and Computation. Business Expert Press, Manhattan, New York, USA, Mar. 2018, Ch. 3, pp. 45–88, in: Huong Ha (ed.), Climate Change Management: Special Topics in the Context of Asia, ISBN: 978-1-94784-327-1, in: Robert Sroufe (ed.), Business Expert Press Environmental and Social Sustainability for Business Advantage Collection, ISSN: 2327-333X (collection, print).

[10] "Multilingual Universal Decimal Classification Summary," 2012, UDC Consortium, 2012, Web resource, v. 1.1. The Hague: UDC Consortium (UDCC Publication No. 088), URL: http://www.udcc.org/udcsummary/php/index.php [accessed: 2020-01-12].

[11] "Creative Commons Attribution Share Alike 3.0 license," 2012, URL: http://creativecommons.org/licenses/by-sa/3.0/ [accessed: 2020-01-12], (first release 2009, subsequent update 2012).

[12] "Perl Compatible Regular Expressions (PCRE)," 2019, URL: https://www.pcre.org/ [accessed: 2020-01-12].

[13] "The Perl Programming Language," 2019, URL: https://www.perl.org/ [accessed: 2020-01-12].

[14] "UDC Summary Linked Data, Main Tables," 2018, URL: https://udcdata.info/078887 [accessed: 2020-01-12].

[15] "UDC, Common Auxiliary Signs," 2019, URL: https://udcdata.info/078885 [accessed: 2020-01-12].

[16] "UDC 9: Geography. Biography. History," 2019, URL: http://udcdata.info/068076 [accessed: 2020-01-12].

[17] "UDC 5: Mathematics. Natural sciences," 2019, URL: http://udcdata.info/025403 [accessed: 2020-01-12].

[18] "UDC "...": Common auxiliaries of time," 2019, URL: http://udcdata.info/011472 [accessed: 2020-01-12].

[19] "UDC (1/9): Common auxiliaries of place," 2019, URL: http://udcdata.info/001951 [accessed: 2020-01-12].

[20] C.-P. Rückemann, R. Pavani, B. Gersbeck-Schierholz, A. Tsitsipas, L. Schubert, F. Hülsmann, O. Lau, and M. Hofmeister, Best Practice and Definitions of Formalisation and Formalism. Post-Summit Results, Delegates' Summit: The Ninth Symp. on Adv. Comp. and Inf. in Natural and Applied Sciences (SACINAS), The 17th Int. Conf. of Num. Analysis and Appl. Math. (ICNAAM), Sept. 23–28, 2019, Rhodes, Greece, 2019, DOI: 10.15488/5241.

[21] A. Bäck, Aristotle's Theory of Abstraction. Springer: Cham, Heidelberg, New York, Dordrecht, London, 2014, ISBN: 978-3-319-04758-4, ISSN: 1879-8578, The New Synthese Historical Library, (Book Series), Texts and Studies in the History of Philosophy, Volume 73.

[22] L. Učník, I. Chvatík, and A. Williams, The Phenomenological Critique of Mathematisation and the Question of Responsibility: Formalisation and the Life-World. Springer, 2015, ISBN: 978-3-319-09827-2, (Collection), Contributions to Phenomenology, Volume 76.

[23] W. H. S. Jones, Pausanias Description of Greece. London: William Heinemann, New York: G. P. Putnam's Sons, MCMXVIII, 1918, vol. I and II.

[24] A. Michaelis, Der Parthenon. Leipzig, Druck und Verlag von Breitkopf und Härtel, 1871, (title in English: The Parthenon).

[25] M. A. Tiverios, "Observations on the East Metopes of the Parthenon," American Journal of Archaeology, vol. 86, no. 2, 1982, pp. 227–229.

[26] K. A. Schwab, Celebrations of Victory: The Metopes of the Parthenon. Cambridge, Cambridge University Press, 2005, pp. 159–198, in: Jenifer Nils (ed.), The Parthenon: From Antiquity to the Present.

# A Data-Driven System for Probabilistic Lost Person Location Prediction

Nathaniel Soule
*Raytheon BBN Technologies*
Cambridge, USA
e-mail: nate.soule@raytheon.com

Stephen Anderson
*Metron Scientific Solutions*
Reston, USA
e-mail: anderson@metsci.com

Colleen T. Rock
*Raytheon BBN Technologies*
Cambridge, USA
e-mail: colleen.rock@raytheon.com

Benjamin Toll
*Raytheon BBN Technologies*
Cambridge, USA
e-mail: ben.toll@raytheon.com

John Ostwald
*Raytheon BBN Technologies*
Cambridge, USA
email: john.ostwald@raytheon.com

James R. Milligan
*US AFRL*
Rome, USA
e-mail: james.milligan.2@us.af.mil

Matthew Paulini
*US AFRL*
Rome, USA
e-mail: matthew.paulini.1@us.af.mil

David Canestrare
*US AFRL*
Rome, USA
e-mail: david.canestrare.1@us.af.mil

James Swistak
*Peraton*
Rome, USA
e-mail: swistak@peraton.com

Eric Daniels
*Peraton*
Rome, USA
e-mail: edanie06@peraton.com

***Abstract*— Today, when a report of a lost person occurs, both the Search And Rescue (SAR) team and Lost Person (LP) have limited access to assistive technologies, leaving manual or ad-hoc search planning as an all too common solution. Geospatial data exists, however, that when coupled with appropriate models and algorithms can enable decision support systems to help predict the location of lost persons and provide guidance for optimal search execution given the available search resources. The environments and context for application of these technologies, however, introduce several key complexities. The data required for accurate analysis and prediction (e.g., elevation, land cover, exclusion zones, known markers) can be large and the exact subset needed for any particular incident may not be known until the lost person event occurs. The algorithms required to generate location probability distributions are compute intensive in comparison to the limited compute resources available on the devices located closest to the incident or carried by a search team. That search team is by design, distributed, conducting operations with multiple independent operators, often in areas with limited, degraded access to network infrastructure. This paper describes the design, algorithms, models, and evaluation of software entitled LandSAR that employs geospatial datasets and tooling in a distributed context to address these challenges and enable such capabilities at the network edge.**

***Keywords-search and rescue; geospatial algorithms; team awareness kit; geospatial data.***

## I. INTRODUCTION

Time and situational awareness are crucial to search and rescue efforts. While there is a plethora of geospatial data to assist and guide action in response to LP incidents, the ability to gather, process, disseminate, and leverage this information is one more challenge at a time of significant risk and stress. Weather, sustenance requirements and injuries all impose a time clock on the search teams. Today's practices are laden with manual elements and thus can only operate at human speed, accuracy and scale, and further require expert knowledge of terrain, personnel and other factors. The work described in this paper presents a machine-speed and machine-scale solution to these issues, with the goal of saving lives, and reducing the duration and thus cost of searches. The LandSAR technology provides a tool that presents probabilities of lost person locations over time, and based on this information, presents search teams with search recommendations given their available assets. Figure 1 depicts this tool executing within an Android-based team Situational Awareness (SA) tool.

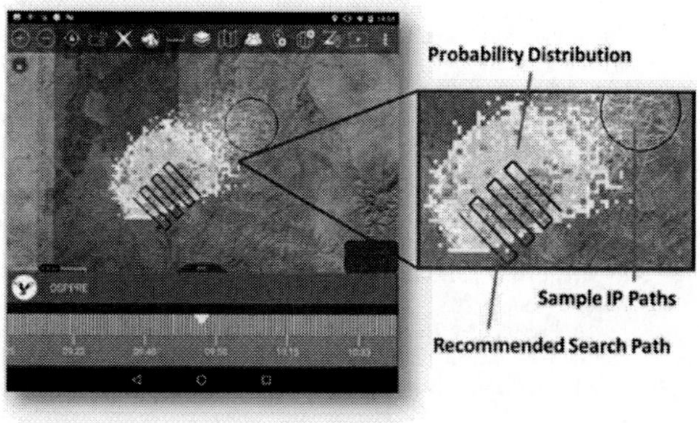


Figure 1. LandSAR UI showing probability distribution, sample paths used by underlying calculations, and recommended search path.

To understand the LandSAR concept of operations, let us consider an event that leads to a lost person who must be located and recovered. LandSAR operators use the LandSAR client software resident on their mobile device, or in a web browser, to record the event and last known position of the LP along with several other parameters, including selection of a model representing the class of LP (e.g., a hiker, someone with dementia), that can help guide the search process. The LandSAR models further capture awareness of destination goals, the most likely selection among multiple such points, and the subsequent choice the LP must make to determine a route to that objective. As an example, a model and its parameters might indicate that the LP has one of several known locations that they may try to get to if lost, and that there may be an area that they will likely avoid if they encounter it (e.g., due to recent flooding that area is no longer easily traversable). The LP will have to make a

Distribution Statement A: Approved for public release; distribution unlimited. AFMC 88 ABW Public Affairs Case Number 88ABW-2019-2275.

This work is sponsored by the Air Force Research Laboratory (AFRL) under AFRL Contracts FA8750-16-C-0116 and FA8750-19-C-0021.

quick rough estimate of the difficulty of such a path based on the information at hand. That information centers on elevation and major features such as water bodies and rivers. Paths which reduce the total elevation deltas and traverse the most amenable land cover types are favored by the LP and thus are also favored by the model. Of course, the LP will not follow the planned route exactly. While keeping with the general trend of the path, the LP will make the final determination based on local factors. For example, the LP will favor open fields over wooded areas to increase visibility. The model will similarly estimate LP paths by using land cover data to make the final determination for each considered path. LandSAR uses these models and accompanying algorithms to develop a time parameterized probability distribution which is sent to the search team, as seen in Figure 1. They can then drag within the interface forward and backward in time to estimate where the LP was, is, or will be. A user can then request that LandSAR calculate an optimal search rectangle and representative search path, providing the system with information about the available assets (searchers on foot, helicopters, or small unmanned aerial systems). LandSAR will generate a recommended search path and disseminate this to the client devices for inspection and execution.

The remainder of this paper is organized as follows: Section II describes related work. The LandSAR system design and its subcomponents are described in Section III. Section IV details the challenges and solutions in geospatial data acquisition and fusion to address the LandSAR information requirements. Section V describes the probability distribution generation and search recommendation algorithms and the models that support them. Section VI describes evaluation of performance and exercise-based evaluation of efficacy. Finally, in Section VII, we conclude and discuss ongoing and future work extending, enhancing, and augmenting the LandSAR capabilities.

## II. Related Work

Search Theory [1][2], a mathematical approach to the search for objects, dates back to World War II and the search for German U-boats. Work in the application of this theory to search and rescue [3], and in particular land-based search, provided the basis on which LandSAR realizes optimal search recommendations.

Other systems exist to predict locations of lost persons and provide search recommendations. SAROPS [4] is a US Coast Guard tool that produces probability distributions using a particle filter for the location of the search object and that recommends search allocations to maximize the increase in probability of detection with the assets available. SAROPS applies these techniques to the sea domain, as opposed to the land-based domain employed in LandSAR.

The Android Team Awareness Kit (ATAK) [5], described in more detail in Section III, is a platform in which the LandSAR capabilities are exposed (in addition to a web-based version). ATAK provides a plug-in interface that allows for easy extension. Other ATAK plug-ins [6] have been developed that project movement to predict potential locations of an entity, but require non-trivial terrain pre-processing and don't provide search recommendations. LandSAR requires only lightweight processing of input datasets before they can be used, and can perform this processing at runtime, is focused more on LP models than determining concealed routes to be used, and recommends optimal search paths based on the determined probability distributions.

## III. The LandSAR System

LandSAR, as seen in Figure 2, is a distributed system where the core computation executes on the server, and the results can be disseminated to a team of users. In this section, we briefly describe these technologies and then provide an overall system view of the LandSAR software.

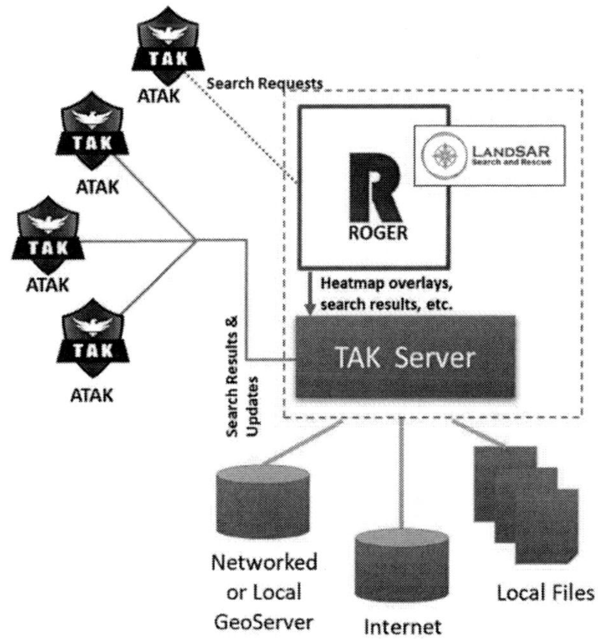


Figure 2. LandSAR Systems View.

ROGER [7][8] is a framework for building modular network middleware by composing plug-ins. The LandSAR capabilities are realized through ROGER plug-ins that model the movement of lost persons over time and provide optimal search recommendations. These plug-ins embody the logic of the search algorithms and work alongside another set of plug-ins that ingest inputs for these algorithms from client devices and local or remote data stores, and a 3rd set of plug-ins that expose the LandSAR outputs to a situational awareness platform called the Team Awareness Kit (TAK). TAK provides a suite of mobile mapping and SA applications employed by over 100,000 US users from numerous local, state, federal, and military agencies. ATAK, the Android-based primary TAK client, supports mobile teams and the wide variety of operating environments and roles that mobile scenarios demand. A server component, called TAK Server, acts as a publish-subscribe middleware connecting ATAK (and other) devices. As shown in Figure 2, client devices communicate with TAK Server for normal SA operations, and can send lost person notifications and search requests directly to ROGER (the box labeled R in the figure) for processing. Generated probability distribution and search recommendation map overlays are distributed to TAK Server as Keyhole Markup language Zipped (KMZ) files to be distributed to all members of the search party, and rendered on ATAK.

The LandSAR client-server model is augmented by the use of a peer-to-peer information management capability called BANDIT [9]. BANDIT processes geospatial situational awareness messages like TAK Server, but does so in a decentralized manner using a light-weight quality of service aware broker on each node in a set of devices. Through the use of mesh networks and the BANDIT technology, LandSAR operations can extend beyond direct line of sight ranges and can handle partitioned group operation (e.g., a subset of searchers out of range of the server or other users).

## IV. Geospatial Data Acquisition and Dissemination

LandSAR requires a number of data inputs in order to generate high quality probability distributions and search recommendations. Some of these inputs may be prepositioned on the server for targeted use cases, while others are required at the time of an lost person event. Several of the key inputs are depicted in Figure 3. On the client side, the last known position

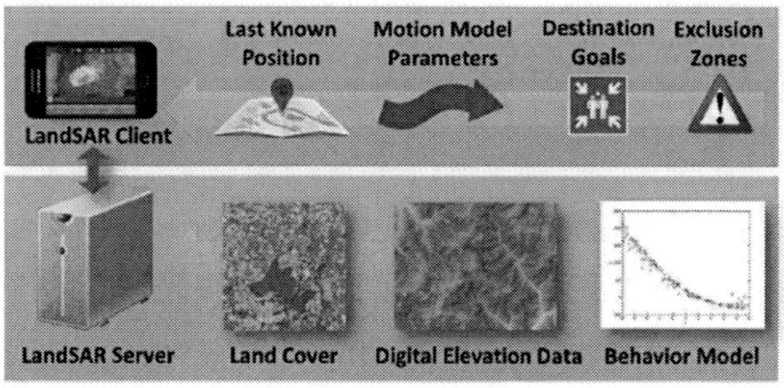


Figure 3. LandSAR Input Data.

is used to center a search. Motion model parameters describe properties of a lost person's movement that may be context specific. For example, a hiker lost in a national park may only move during the day time. Destination goals describe locations known to the LP that they may be more likely to move towards. Exclusion zones are those places an LP is more likely to avoid, and fall into two categories: those known in advance, and those that may be discovered during an event. On the server side the datasets become larger. Land cover data describes the type of terrain (e.g., deciduous forest, grasslands) and is used to help calculate the Speed Of Advance (SOA) – how fast someone may move across that terrain. Digital elevation data is similarly used to determine a realistic SOA and to guide path selection (e.g., an LP may favor flatter terrain over mountainous). Behavior models are used to guide estimation of how an LP will act and make decisions while lost (these models are discussed in detail in Section V.A). These inputs present challenges in terms of both data hygiene, and data acquisition, dissemination, and storage at the tactical edge.

The accuracy and precision of the LandSAR algorithms is a direct function of the quality of the data inputs. Many elevation data sources, for example, contain voids – spaces for which no sensor data was present in the dataset. These voids are often filled with marker values (e.g., some extreme minimum or maximum) and can impact path prediction. An extreme positive value used as a void-filler, for example, might lead the algorithms to believe extreme elevations are present in a path when in fact none exist. Smoothing operations, or dataset fusions to fill such gaps are thus useful. In testing and evaluation, LandSAR thus often uses the Shuttle Radar Topography Mission (SRTM) [12] Version 2 elevation data, which has been post-processed by the National Geospatial-Intelligence Agency. Among other improvements, this post-processing removed single pixel errors and defined coastlines. SRTM data for the United States is accurate to within 1 arc-second, or 30 meters.

While highly accurate data is useful for high quality results, such data also implies large storage requirements and retrieval costs. Each SRTM file, for example, is 25.9 MB once unzipped, and represents 1° latitude by 1° longitude. The 968 such files across six regions of the contiguous United States (CONUS), are, therefore, just over 25 GB in size. All available SRTM data was also obtained in Digital Terrain Elevation Data (DTED) format from the U.S. Army Geospatial Center, consisting of 13,986 files totaling 322.9 GB. For search operations in CONUS, land cover data is obtained from the National Land Cover Database (NLCD), available from the Multi-Resolution Land Characteristics (MRLC) consortium. The CONUS NLCD data is 1.1 GB when compressed in a zip file, and 18.3 GB uncompressed. Visual Navigation (VISNAV) [10] land cover data covers the globe, excluding CONUS and Alaska, and is 17,437 files totaling 288.5 GB. Importantly, the NLCD data and the VISNAV data have the same resolution as SRTM data, allowing for a consistent discretization of the area of interest.

The LandSAR data acquisition design was tailored specifically to address the challenges posed by large datasets representing the areas relevant to a mission, while accounting for the constrained networks, and limited resources of devices at the edge traditional network connectivity. LandSAR allows users to specify an Area of Interest (AOI) within which the LP is likely to be located. Determination of the appropriate area is based on the space an LP could cover within the time period of interest (e.g., a few days). LandSAR also provides support for obtaining elevation and land cover data from a mission/deployment-scoped dataset on a GeoServer [13], an open-source Java-based software server that allows access to geospatial data using open standards. A GeoServer can be co-resident with the LandSAR gateway, or, for deployments with sufficient network connectivity, hosted remotely. LandSAR also has a capability to load data directly from local disk to avoid the need for additional data servers. In cases where high bandwidth Internet connectivity is available, LandSAR can also access data directly from public sources, on-demand as needed, without any preloading.

While LandSAR currently relies heavily on elevation data and land cover data, the algorithms and models can be further tailored to specific scenarios with more data (e.g., weather). Additionally, LandSAR has support for a trails-based motion model, which requires machine-ingestible trail data.

The more data available to LandSAR, the more accurate and precise the results can be. Getting access to additional data feeds where the search teams operate and network connectivity is often limitedcan be challenging; however, work is currently underway to be able to ingest and employ data that is already available at the network edge. LandSAR is positioned within the TAK software suite so that it can consume situational awareness data already flowing through TAK Server, allowing for knowledge of team locations, structures, landmarks, routes, etc., to help refine probability distributions and search results. For example, we are currently exploring probabilistically generated exclusion zones and LP destination goals, based on existing team reported locations, speed and heading data.

Beyond data acquisition and processing, data dissemination presents numerous challenges in the constrained networks in which search and rescue often occurs. The aforementioned BANDIT capability is able to shape the data that flows through each node to meet the constraints of the network. LandSAR enhanced this capability to deal directly with the data formats of concern in SAR operations, such as KMZs which describe place marks, images, polygons, etc. that can be overlaid on a map. Format specific KML compression techniques are being employed by LandSAR. This is accomplished by compressing the points in each line segment described in the KML. Each one of these lists is run through the Microsoft Bing point compression algorithm [11], which generates a single compressed string representing the entire list of points. Each of these strings is then stored in a JSON array and then further compressed using a simple GZip compression. In early test results, a 2MB KML file is initially reduced to 140 KB and after final processing including the GZip step, to 30 KB.

The size of the data dissemination is only one component of addressing effective dissemination in search and rescue contexts. Radio compatibility is another concern. A search team's effectiveness is in part a function of its size, and thus allowing for ad hoc team augmentation is desirable in some scenarios. The team members added in this fashion, however, may not have devices with compatible radios. LandSAR is thus using a QR code transmission mechanism that allows sharing search recommendations using only the camera and screen of mobile devices. QR codes are ubiquitous for visually transferring small amounts of data without a network connection, but their bandwidth is limited, especially by the displays and cameras of mobile phones. The BANDIT technology provides a streaming QR code capability that can be thought of as a flip book of QR codes. This went a long way to mitigate bandwidth limitations by allowing data to be transferred through multiple QR codes, but even this has limits, as it requires the sender and receiver to be physically still, and in close proximity for extended time periods. To achieve even greater bandwidth LandSAR is being extended with the use of color in QR codes. Using color allows more data to be stored in each pixel, thereby increasing their bandwidth and shortening data transfer time. As an example, using 16 colors in a QR code, instead of the normal 2 (black/white), would allow a four-fold increase in bandwidth, essentially quartering the time required to send the same amount of data. It is not without difficulty, though, as introducing color increases the computation complexity of encoding / decoding the data, and it introduces another source of possible errors which must be dealt with, especially in a mobile setting where local lighting conditions can vary.

## V. Algorithms and Models

### *A. Modeling the Lost Person's Location*

In typical lost person events, the search team decision makers have too many unknown variables, limiting their confidence of where to search. LandSAR attempts to help the decision maker by modeling possible outcomes of where the LP could be over time using different combinations of these variables in conjunction with land cover, elevation and other data to estimate how fast and in what direction the LP might travel.

The system generates a path based on several interrelated models and the available elevation and land cover data, beginning from an initial distribution of potential starting points. It will repeat this procedure many times to produce a sufficiently representative collection of possible paths for the LP. Given this set of paths, LandSAR can provide an estimate of where the LP will be at any time in the future. The simulation ultimately generates a heat map that visually depicts the probability distribution of the LP at any given point in time. An example of this output is shown in Figure 4. On the heat map, red equates to 50% probability that the LP is in these rasters. If you combine red and orange, then 90% of the generated LP paths are in these rasters. If you combine yellow on top of orange and red, then you have all possible outcomes of where the LP was predicted to be according to the simulation.

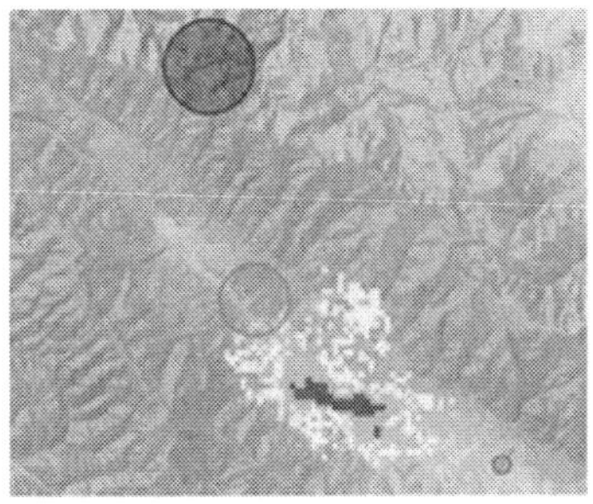

Figure 4. LandSAR generated probability distribution heatmap.

LandSAR uses three model types to probabilistically estimate the location of the LP: a model of where the person starts, a motion model capturing decision processes and ultimately impacting the paths they could traverse, and a model of how fast they traverse those paths. These are described in detail in turn below.

#### *1) Modeling an LP's Starting Point and Intial Movement*

To understand where an LP may be at a future point in time, it is critical to understand where they may have been at some point in the past. An exact time and location may not always be known, and thus the starting point model often has incomplete information. There may, therefore, be a distribution of possible starting locations based on the last received information from the LP. LandSAR allows the user to represent these initial distributions via a uniform circle, uniform polygon or several other methods.

Figure 4 shows the likely location distribution from a lost person moving towards one of several potential rendezvous points (one shown as a green circle) and avoiding a known exclusion zone (red) and discovered exclusion zone (orange).

#### *2) Motion Models*

LandSAR motion models estimate what an LP is likely to do by integrating assumptions about the thought processes of the LP with information about the area in which they are isolated. A LandSAR user chooses a motion model to best fit the circumstances of the LP. The motion model is used to produce a probability distribution for the location of the LP over time. The approximation for this distribution is a set of sample paths. There is a tradeoff between the number of sample paths (and thus the quality of the results) and computational cost. The more sample paths available to represent the distribution, the better the approximation. The computational cost scales linearly in the number of sample paths. There is, of course, significant uncertainty in that process and, consequently, a Monte Carlo technique is applied to determine the probability distribution of where the LP could be over time.

TABLE I. EXAMPLE LANDSAR MODELS.

| Model | Description |
|---|---|
| **Stationary** | The LP is assumed to be injured |
| **Lost Hiker With Destination** | The LP knows where they are and where they must go. They move in the terrain that best affords success in reaching their goal. |
| **Trails-Based** | The LP will move until they reach a trail and then follow it in one direction until found. |
| **Easiest Short-Term Path** | The person does not know where they are nor do they have an idea of where help may be, and will take the easiest short-term path |

LandSAR supports a number of motions models, examples of which are listed in TABLE I.

*3) Speed of Advance*

The generated sample paths will describe where an LP may go, but not when they will be there. Their speed of advance along each path is needed to account for the temporal dimension. That speed will be a function of the steepness of the path, the type of land cover, the physical fitness of the LP and other aspects. The Speed of Advance (SOA) model includes all of these factors. After LandSAR determines an initial route that only takes into account water features and elevation variation, it will then utilize other costs to determine minor variations from this route. As LandSAR forms a feasible route, it moves through land cover which has the lowest cost available to choose from. A user may choose to change the default costs to account for assumed choices the LP would make. In the case of an LP moving through woods, for example, the user may assume he/she would choose to move through less densely wooded areas to provide better visibility and thus the user would change the values of deciduous, evergreen and mixed forests to be less likely to be considered. The adjustment factor for slope is a modified version of the formula for walking speed adjustment based on slope, also known as Tobler's Hiking Function [14]. The SOA models take into account facts such as a gentle downslope increases speed of advance, while a steep downslope decreases it.

Finally, a user can also set a movement schedule for the model of the LP, if for example, the LP is likely to only move during the daylight hours, or in some cases, only at night.

## *B. Search Optimization*

The estimate of location in the future can be used to aid attempts to rendezvous with the LP. Maximizing the probability that the LP is localized uses elements of search theory such as the Koopman random search formula [15]. A probabilistic technique is used to find an appropriate search plan. The appropriate plan will, of course, depend on the capabilities of the search asset. A helicopter can cover more area but a searcher on foot might have better probability of detection. Koopman's random search formula is currently used in most of the search algorithms in LandSAR. A more sophisticated approach using lateral range curves, similar to that employed in the U.S. Coast Guard's search application SAROPS, could be readily implemented. However, the sensors used for land-based search and the environmental conditions affecting their performance have not been successfully modeled to the same level as those used for maritime searches. Consequently, we employ the simpler formula, which takes into account the speed, height above ground, and sweep width for the asset. Obviously, the longer an asset can stay on station and search the area, the higher the probability of detection.

LandSAR provides a search area with an associated Probability Of Success (POS), given the search assets and the time they can search for the LP. To do so, LandSAR generates 1,000 random search boxes and calculates the probability of success for each search box. It will recommend the search box with the highest POS and then make small adjustments (e.g., offset, rotate) to improve it. The POS is defined as $P(\text{success}) = P(\text{containment}) * P(\text{detection})$, where the probability of containment is the likelihood of the search object being contained within the boundaries of some area. It is possible to achieve 100% POC by making the area larger and larger until all possible locations are covered, though data and computation requirements scale as the area considered scales. Probability Of Detection (POD) is the likelihood of detecting an object or recognizing the search object and the POD generally decreases the farther away the asset is from the target. It is assumed that the search object is stationary during the search. As long as the search duration is no more than a few hours or the distribution is no longer changing with time, this is a reasonable approximation consistent with the level of detail elsewhere in the modeling. Searches of longer duration can be broken into shorter time intervals to account for these constraints.

# VI. EVALUATION

Our evaluation of LandSAR, like most evaluations, considers both efficacy and performance/resource-cost. Realtime access to isolating events, which happen at unpredictable and stressful times, makes efficacy difficult to measure. Below, we present a small instance-based efficacy evaluation based on use of LandSAR algorithms in exercises and in real LP events where the system was used in parallel to existing manual methods as a way to judge early efficacy without yet completely relying on the system (and potentially putting lives at risk). We next measure the performance of LandSAR to understand how fast it can execute (time is often of the essence in SAR operations) and what device resources it requires for complete and performant operation.

## *A. Efficacy*

Multiple LandSAR evaluations showed the system accurately predicting the location of lost persons in the areas that they were actually found. The number of available lost person incidents during the evaluation were insufficient to provide a true statistical basis, but provided enough evidence of efficacy to warrant subsequent evaluations which will begin in the spring of 2020. Though successfully finding a lost person is the ultimate goal, LandSAR has the secondary effect of reducing the man and flight hours committed to searches. In 2018, the U.S. Air Force Rescue Coordination Center (AFRCC) reported that they responded to 933 SAR mission and the CAP flew 752 missions. Any reduction of time across so large a number of missions has the potential to save lives and to significantly reduce costs for the respective government agencies.

### B. *Performance*

We measured runtime performance across multiple devices, looking at how area of interest size and model selection impacted overall execution. Here, we report experiments run on an Intel® Xeon® Dual 4-core laptop with 32GB of RAM.

TABLE II. EVALUATION GEOGRAPHIC REGIONS

| Area Name | Area Description | Center Point (lan/lon) |
|---|---|---|
| **NM** | New Mexico / Arizona border | 32.0, -109.1 |
| **MA** | Massachusetts | 42.187279, -73.005823 |
| **MI** | Michigan | 44.017543, -84.252951 |
| **NW** | Near Coeur d'Alene National Forest, Idaho | 47.75, -116.6 |
| **RockyMs** | Rocky Mountains | 44.268656, -109.786399 |

Five geographical areas, listed in TABLE II, were analyzed. For each area, the center point was used as the single starting point for the LP and bounding boxes of three different sizes were considered:

**Large**: 107 km east to west by 125 km north to south
**Medium**: 35.8 km east to west by 36.6 km north to south
**Small**: 12.0 km east to west by 14.0 km north to south

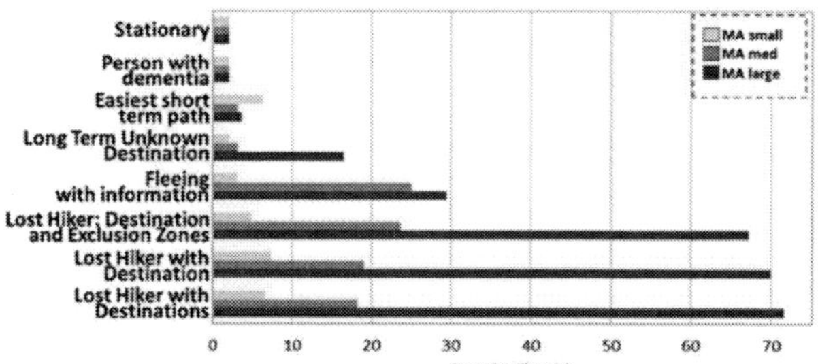


Figure 5. Execution Times for Specific Motion Models.

Figure 5 depicts the overall runtime of the system when operating on the MA small, medium, and large data sets for each of 8 LandSAR models. As can be seen, simple models, such as the stationary model, operate quickly regardless of dataset size. More complex models, such as those that must consider rendezvous points or exclusion zones are more dependent upon the geographic area size (and thus the data size). In the slowest configuration – the large dataset with the most complex model, the runtime is just over 70 seconds, and thus still a very feasible duration for real world contexts.

Figure 6 and Figure 7 look at CPU and heap memory usage respectively, when executing over the MA large dataset. CPU usage, after an initial ramp up, consumes the vast majority of the system's compute resource, indicating multiple concurrent search executions, or slower processors could likely lead to meaningful increases in overall runtime.

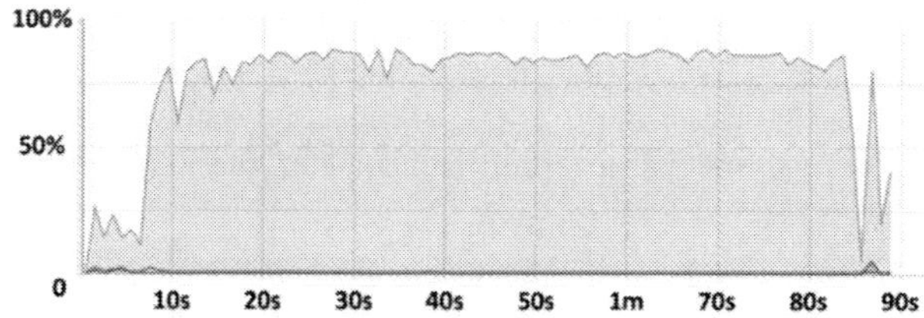


Figure 6. LandSAR CPU Usage on MA Large Dataset.

It can be seen that while the operations employ a non-trivial amount of RAM (peaking around 2GB), it did not come close to consuming the 32GB of available memory on the machine (here we show heap memory; non-heap memory usage was low, approximately 25MB).

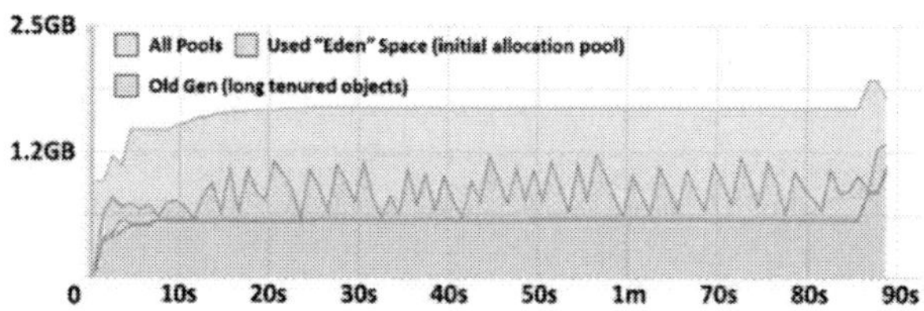


Figure 7. LandSAR Heap Memory Usage for the MA Large Dataset.

## VII. Conclusions and Future Work

The LandSAR capabilities described here have shown initial promise in both performance evaluation and in early trials. Embedding of this technology into the TAK platform enables both increased evaluation and increased likelihood that the capabilities will be in the hands of those that need them when and where they are needed. This work is forming a base on which a set of optimized, enhanced, and augmented capabilities are being built. Deployment in real SAR contexts is underway, and work is being undertaken in a number of areas of technical and capability advancement:

- Enhanced accuracy and precision through the ingestion of situational awareness data that is already natively flowing through TAK devices.
- Development of a web-based version to support search command centers and teams without ATAK devices.

- Automated and semi-automated tasking of small unmanned aerial systems (sUAS) based on LandSAR-generated search recommendations.
- Employing streaming color-coded QR codes for increased bandwidth when sharing search information with joint or other forces that may not have compatible radios.
- Extending the LandSAR format-centric compression techniques tailored at reducing size of the KMZ files through the use of point reduction algorithms such as [16] to decimate a curve composed of line segments to a similar curve with fewer points.

This suite of capabilities, combined with the current LandSAR functionality, will result in a SAR- and LP- focused tool that has the potential to dramatically reduce the duration of LP events, and increase the likelihood of successful rescue operations.

## REFERENCES

[1] L. D. Stone, Theory of optimal search, vol. 118. Elsevier, 1976.

[2] B. O. Koopman, Search and screening: general principles with historical applications Vol. 7. New York: Pergamon Press, 1980.

[3] J. R. Frost and L. D. Stone, "Review of search theory: advances and applications to search and rescue decision support" (No. CG-D-15-01). Soza and Company LTD Fairfax VA, 2001.

[4] T. M. Kratzke, L. D. Stone, and J. R. Kratzke, "Search and rescue optimal planning system," In 2010 13th International Conference on Information Fusion, IEEE, July 2010, pp. 1-8.

[5] K. Usbeck et al., "Improving situation awareness with the Android Team Awareness Kit (ATAK). In Sensors, and Command, Control, Communications, and Intelligence (C3I) Technologies for Homeland Security, Defense, and Law Enforcement" XIV Vol. 9456, p. 94560R, International Society for Optics and Photonics, May 2015.

[6] Ground Guidance Data Sheet, http://www.primordial.com/documents/ground_guidance_military_datasheet.pdf, accessed April 16, 2019

[7] N. B. Soule et al., "Enabling real-time global reach using a gateway building framework," In MILCOM 2017-2017 IEEE Military Communications Conference (MILCOM), pp. 592-598.

[8] C. T. Rock et al., "Efficiently composing validated systems integration gateways for dynamic, diverse data," In MILCOM 2019-2019 IEEE Military Communications Conference (MILCOM), pp. 268-275.

[9] N. D. Holzhauer, J. R. Milligan, and N. B. Soule, "A hybrid P2P and pub/sub messaging system for decentralized Information Management," In MILCOM 2016-2016 IEEE Military Communications Conference pp. 1016-1021, November 2016.

[10] J. M. Wilhoit, N. R. Myers, and G. W. Calfas, "Data collection and management with ENSITE HUB: ENSITE HUB version 1.0," No. ERDC/CERL SR-17-14, ERDC-CERL Champaign United States, 2017.

[11] Point Compression Algorithm – Bing Maps: https://docs.microsoft.com/en-us/bingmaps/rest-services/elevations/point-compression-algorithm. Accessed April 16, 2019.

[12] T. G. Farr et al., "The Shuttle Radar Topography Mission", Rev. Geophys., 45, RG2004, 2007, doi:10.1029/2005RG000183.

[13] S. Iacovella, GeoServer Beginner's Guide: Share Geospatial Data Using Open Source Standards. Packt Publishing Ltd, 2017.

[14] T. J. Pingel, "Modeling Slope as a Contributor to Route Selection in Mountainous Areas," *Cartography and Geographic Information Science* 37.2, 2010: 137-148. Web. 2015.

[15] J. R. Frost. The theory of search: a simplified explanation. Soza Limited, 1997.

[16] A. Saalfeld, Topologically consistent line simplification with the Douglas-Peucker algorithm. Cartography and Geographic Information Science, 26(1), p 1999.

# Electric Energy Consumption Forecast Based on Spatial Information

Carolina L. S. Cipriano *, Mayara G. Silva*, Weldson A. Corrêa*,
João D. S. Almeida*, Márcia I. A. Silva†, João O. B. Diniz*
*Applied Computing Group (NCA), Federal University of Maranhão (UFMA), São Luís - MA, Brazil
Email: {carol, mayara, weldson, jdallyson, joao.bandeira}@nca.ufma.br
†Equatorial Energia, Brazil
Email: marcia.alves@equatorialenergia.com.br

*Abstract*—**The task of predicting the consumer's electricity consumption is currently a trend in power energy companies. This prediction becomes difficult or impractical for consumers with no history or a short history of consumption. Thus, this work deals with an alternative to the prediction of energy consumption for these consumers. The proposed method is based on the consumption of the k closest neighbors and the consumption forecast made by one of three available regression models. The regressors used, namely Autoregressive Integrated Moving Average (ARIMA), Boosting Additive Quantile Regression (BAQR) and the named Seasonal and Trend decomposition using Loess (STL), were chosen for providing the best performance. The results obtained were promising, achieved a mean of the 30.4 % in the symmetric mean absolute percentage error (sMAPE) metric in a dataset with 86,874 customers.**

***Keywords–Geospatial Information; Energy Forecasting; STL; ARIMA; BAQR***

## I. Introduction

It is a current trend for power companies to invest in artificial intelligence and machine learning to predict the monthly behavior of their consumers' energy consumption [1][2]. Forecasting is beneficial for both energy companies and consumers. This mutual benefit comes from reducing the energy company's expenses during power distribution and increasing its revenues. By reducing financial losses caused by wrong measurements or power thefts, it can then pass on to its consumers a lower energy consumption billing rate.

The problem of predicting consumers' electricity consumption is an essential step in verifying inconsistencies in measuring monthly energy consumption. Inconsistency checking avoids both incorrect billing for a consumer and may indicate that, due to abnormal energy consumption, the consumer may be using technical arrangements to reduce his energy consumption and, therefore, not being properly registered. For this reason, power companies have invested in Pattern Recognition (PR) methods to predict their customers' energy consumption and thus improve the verification step for measuring energy consumption inconsistencies.

In practice, each company defines its criteria for checking for inconsistencies in reading data. For this verification, Equatorial Energy uses the average consumption of the last threes months as a forecast for each consumer. Forecasted consumption is used to define a minimum and maximum consumption range. This range is defined to avoid errors and anomalies in the reading performed. Readings outside the expected range are reviewed by company technicians before issuing the customer invoice.

Our motivation stems from the fact that this task of predicting the behavior of power consumption is relatively simple when its consumers' consumption history exists, but it becomes difficult or impractical for consumers with no consumption history, i.e., for new consumer installations or those who have a short history of power consumption. Consequently, the reading of these customers usually goes through technical analysis before issuing the invoice, given the impossibility of predicting consumption.

In this sense, to solve the problem of energy consumption prediction of consumers without consumption history and reduce the number of customers that are analyzed before the invoice issue, this paper proposes an energy consumption prediction method using neighborhood consumption information. Indeed, it is likely that a new customer will have a consumption similar to that of its closest neighbors, as well as its consumption range.

In the proposed work, the term spatial information is related to neighborhood identification, to compute energy consumption based on Tobler's first law of geography [3] stating that everything is related to everything else, but near things are more related than distant things.

This work is part of a Research and Development (R&D) project, contracted by Equatorial Energy under contract CELPA 962/2018 and CEMAR 30/2019, executed by the Applied Computing Group (NCA) of the Federal University of Maranhão (UFMA). This project will develop the Consumption Habit Analysis System (SisHCo). The project is organized to provide the development of methods, techniques, and tools based on computational intelligence and machine learning, to define, in an individualized and adaptive way, parameters for the critique of power consumption measurement, based on historical information.

The main contributions resulting from this work are:

1) We developed a method that solves a real problem of the energy supply companies, using simple techniques and with reasonable accuracy;
2) An alternative method was developed to forecast energy consumption in customers without a history of consumption;
3) The proposed method uses spatial data as a step in the forecast flow of energy consumption;
4) We determined and compared the most accurate prediction methods.

The rest of the paper is organized as follows. In Section II, we present the main related works. Section III describes the proposed methodology of the prediction of individual power consumption of consumers. The presentation of the results is given in Section IV. Finally, a conclusion on the results obtained is drawn in Section V.

## II. Related Work

A Long Short-Term Memory (LSTM) network was used by Alonso et al. [4] for predicting the individual hourly load data of consumers. The prediction model was generated from consumption, weather and calendar data. In the construction of the model, data from 3,891 smart meters in 2013 were collected, producing 8,760 readings on each meter. Then, it was evaluated how the spatial location of the residential customers influences the load prediction. Results indicate that the proposed model mean-absolute error was 19 % better than that of the ARIMA [5] and around 24 % better than that of the seasonal naive approach [6].

Bâra and Oprea [7] evaluated the dynamic profile of energy consumption of consumers with and without power generation. Their objective was to develop a Neural Network (NN) for predicting energy demand based on smart grid consumption patterns and profiles. For this purpose, the profiles were created using a Self-Organizing Map-based pooler (SOM) and an autoregressive NN for daily prediction of energy consumption. As for the power generation properties, a feed-forward NN was used to predict the consumption. Results were obtained in a dataset consisting of 212 consumers, with approximately 1,900,000 hourly energy consumption registers from several devices. Clustering with SOM produced better results than with k-means. The consumption prediction of consumers without power generation resulted a correlation coefficient of 0.99429, Mean Squared Error (MSE) of 0.0046 and a Mean Absolute Percentage Error (MAPE) of 4.21%. Similar results were observed for consumers with power generation, in which the correlation coefficient was 0.999 and the MSE was 0.04.

Jiang et al. [8] proposed a fuzzy clustering model to categorize consumers and identify their energy consumption characteristics. The identification of such characteristics is done after each customer's consumption series are individually grouped into similar parts to detect consumption patterns. Then, the fuzzy clustering generates groups of customers with the same consumption profile, from which the features of consumption patterns are extracted. Finally, a classifier is used to categorize new consumers into one of the previously found groups. Results were obtained for a dataset of hourly energy consumption from 1,168 non-residential consumers over a year. The authors concluded that their fuzzy clustering-based method improved classification accuracy for the inclusion of new consumers.

The prediction of daily energy consumption in apartments in the Republic of South Korea appears as a problem in the work of Wahid and Kim [9]. In this work, K-Nearest Neighbors (KNN) was used as a predictor of energy consumption over hourly consumption data from 520 apartments. From the consumption history, four features were extracted: average, variance, asymmetry and kurtosis. An accuracy 95.96% was obtained as best result.

Lora et al. [10] compared the energy price time series prediction performance of two models, one based on a multilayer perceptron recurrent neural network, and the other based on a combination of KNN and Genetic Algorithm (GA). They used GA to adjust the weights for Euclidean distance. The performances of both models were compared in a small dataset of energy prices from January to August 2001, in which was obtained in the period from March to May an MSE of 0.3464, and in the period between June and August an MSE of 0.428.

Poloczek et al. [11] and Kim et al. [12] used KNN to predict the values lost in the process of sensor data acquisition, due to inactivity. Both works showed that KNN is able to produce results close to real values, using both the proximity of the data values and the sensors' spatial information. These results motivated us to use both the spatial information of energy-consuming facilities and KNN, for its simplicity in data generation.

Most of the aforementioned works use consumption readings automatically acquired on smart meters, which are less sensitive to noise caused by acquisition mistakes. Our proposed work, however, makes use of a dataset which acquisitions were made manually in electromechanical and digital meters and, as a consequence, are subject to mistakes during the readings of energy consumption. Therefore, the presence of noise makes the task of forecasting consumption more challenging.

It is worth mentioning that only a small group of clients use smart meters, which explains the small number of clients in these works. In our work, though, we use a large customer dataset with more than one year of energy consumption data. Moreover, it can be observed that in Jiang et al. [8] there is a need for initial consumption data for new customers, so that they can be inserted in a group. This is necessary to enable the use that group's model in future predictions. We highlight that this restriction is not present in our work.

Additionally, the works of Wahid and Kim [9], Lora et al. [10], Poloczek et al. [11], and Kim et al. [12] perform data prediction using only KNN and spatial information due to its excellent performance in the data regression process. Similarly, our work uses KNN, spatial information, and energy consumption data to estimate new customer consumption. However, it differs from the mentioned studies in that it uses the best prediction result among three proposed regressors, obtained in eight distinct classes of consumers.

## III. Materials and Method

This section describes the materials and the proposed method for consumption estimation of new installations. The steps of the proposed method are presented in the sequence they are applied, as illustrated in Figure 1. First, the process of data acquisition is described. Second, the dataset goes through a preprocessing step. Third, the neighborhood of each customer is determined. Fourth, consumption of the customer's neighbors is predicted. Fifth, consumption of customers with short series is predicted. And, finally, results are validated in step six.

### A. Data acquisition

The dataset consists of power consumption data from 2,316,760 active customers from the state of Maranhão, Brazil. The data was collected monthly from January 2017 to April 2019, and form the series consumption history for each customer.

Customers are organized into classes and subclasses, according to ANEEL Normative Resolution no. 414/2010 [13], repealed by ANEEL Normative Resolution no. 800/2017 [14]. The consumption classes applicable to consumers are:

- **Residential**: this category includes consumer units with residential purposes;

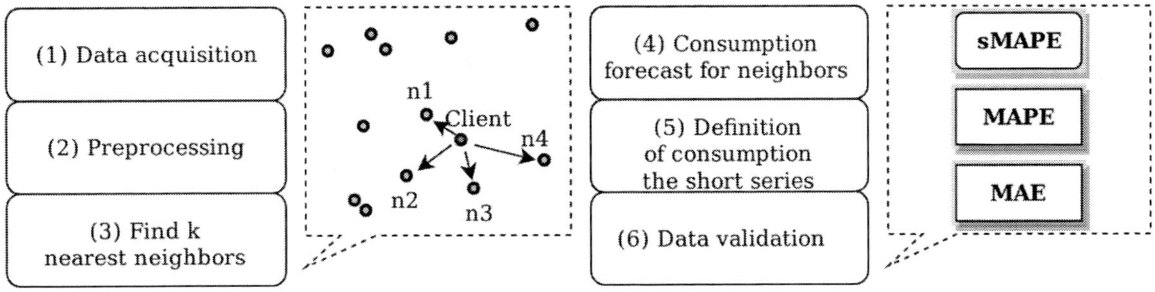


Figure 1. Steps of the proposed method.

- **Industrial**: are the consumer units in which industrial activity is developed;
- **Commercial, services and other activities**: this includes the consumer units where the service rendering activities are developed and others not provided for in the other classes;
- **Public service**: consumer units are intended exclusively for the supply of engines, machinery and cargo essential for the operation of public water, sewage, sanitation and urban or railway traction services, operated directly by the Government or through concession or authorization;
- **Self-Consumption**: the consumer units owned by the distributors are included;
- **Rural**: consumer units that develop activities of agriculture, livestock or aquaculture;
- **Government**: consumer units that are consumers of a legal entity governed by public law are independent of the activity developed, including illumination on roads and traffic lights, radars and traffic monitoring cameras, except for those classified as public irrigation services, schools, agrotechnics, street lighting and public service;
- **Public Lighting**: public service whose sole purpose is to provide clarity to public places on a periodic, continuous or occasional basis.

### *B. Preprocessing*

Before the short-series consumption forecast step, the dataset is subjected to a preprocessing step. In the dataset there are customers with short series, ranging from zero to four months of registered consumption, totaling 95,052. For this reason, this data is separated and used in the tests of the proposed method. In addition, we ignore consumer series that: (1) do not have at least two neighbors to be considered in their estimation, i.e., series categorized in classes that have less than threes installations; (2) cases where clients do not have valid coordinates, which makes it impossible to identify their location and distance to their neighbors; and finally, (3) cases in which the series do not have consumption registered in the reference month of this study, making it impossible to validate the estimated consumption.

### *C. Finding K-Nearest Neighbor*

The k-nearest neighbors are defined based on the customer's geographic coordinates. Fig. 2 shows the information available on each neighbor, which is the prediction of consumption for the reference month of June 2018, the prediction interval with minimum and maximum (after III-D); the coordinates with latitude and longitude; and the Reading Unit (RU). The reading unit represents a set of installations that are read by a particular reader on a reading day. RU information is used to narrow the search scope of K-neighbors. Instead of searching for K-Neighbors throughout the municipality, only the neighbors belonging to the same RU of the analyzed series are searched.

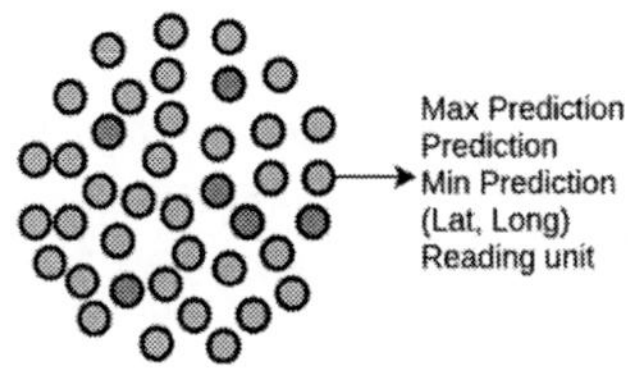


Figure 2. New customers in red and their neighborhood in green.

On the other hand, in the case of a larger number than K-neighbors are available with the same coordinate, only the k-neighbors with the greatest consumption history will be selected.

### *D. Consumption forecast for neighbors*

Consumption forecast and neighbors minimum and maximum prediction interval were performed for the reference month of June 2018. Consumption was estimated using statistical methods (STL [15] and ARIMA [5]) and methods based on machine learning (BAQR) [16]. These methods were empirically chosen based on tests because they outperform most classes over other methods, such as LSTM [17], SGD [18]. STL was better for the Industrial and Self-consumption class, ARIMA in the Public Lighting and Public Power class and BAQR was superior in the Residential, Rural, Public Service and Commercial classes.

*1) Boosting Additive Quantile Regression (BAQR):* It is a quantile regression model that uses additive models to relax the assumption of linearity [16][19]. It is based on the following equation:

$$\hat{q}_\alpha\left(x_1^{(i)}, ..., x_d^{(i)}, z_1^{(i)}, ..., z_J^{(i)}\right) = \beta_0 + \sum_{j=1}^{d} \beta_j x_j^{(i)} + \sum_{j=1}^{J} g_j(z_j^{(i)}) \quad (1)$$

where $g_j$ is a variable smoothing function $z_j^{(i)}$. The method uses the boosting technique to estimate the model by minimizing a loss function using the descending gradient method.

*2) STL:* According to Cleveland et al. [15], STL is a filtering procedure that decomposes the series into Trend ($T$), Seasonality ($S$) and Error ($E$) components. Thus, the original series can be formed by the sum of these components. So, seasonality determines the existence of a cyclic pattern in a time series and the trend is characterized by the behavior of growth or decrease in the long-term amplitude of the time series. Therefore, the power consumption series was filtered using STL and the result was used to choose two regressors: *Simple Exponential Smoothing* (SES) and *Holt's Linear Smoothing*(Holt).

The first regressor, SES [20], shows better results when the dataset has no trend or seasonality, while the second, Holt [21], when there is a trend but no seasonality. Therefore, because these regression methods were input using the result of STL filtering, this method was named STL.

*3) Autoregressive Integrated Moving Average - ARIMA:* It is defined as a generalization of the Autoregressive Moving Average (ARMA) model. These models are generally applied to non-stationary data because, through the differentiation step, this data is transformed into stationary data [5].

### E. Definition of consumption of the short series

This section presents the method for estimating the k-nearest neighbors of the new customer. Consumption estimation for new, short-series, non-consumer customers depends on the discovery of k-nearest neighbors, as well as the consumption estimation and prediction interval of the k-nearest neighbor which is described in the Section III-D.

In the study, the developed methodology considers four scenarios for consumption estimation and prediction interval. The first scenario is made for customers without consumption history and the other scenarios are made for customers with short series. June 2018 is used for validation of results only, not entering each client's series size count. Therefore, customers who have only the June 2018 validation reference month, for example, have zero-size series.

The first scenario considers only customers who have no history of consumption. The second scenario considers customers who have a single month of consumption. The third scenario considers customers who have two months in their consumption history. And finally, the fourth scenario considers customers who have three or four months in their consumption history.

For the first scenario, in which the customer has no history, the consumption estimation and prediction interval are generated from its vicinity, as shown in Fig. 3. In this figure, in (1) are the nearest neighbors of the new customer, all belonging to the same RU, which is defined by the company. In (2), the median of consumption predictions and prediction intervals of the k-nearest neighbor are calculated.

For the second scenario, the approach repeats the previous month's consumption for the prediction of consumption. For the third scenario, the approach calculates the median consumption of previous months to predict consumption. However, the consumption interval estimation of these scenarios continues to be made by estimating the neighborhood consumption interval.

In scenarios two and three, a problem was encountered regarding the generated consumption interval. In some cases, the prediction based on previous customer consumption may be outside the range generated by neighbors. This problem was mitigated as follows: the ratio between the upper limit and the estimated customer prediction found from the neighborhood method is obtained and this factor is multiplied by the customer's predicted consumption, due to its short history of consumption.

### F. Validation of Results

The proposed method was evaluated using the MAPE, Symmetric Mean Absolute Percentage Error (sMAPE) and Mean Absolute Error (MAE). These metrics are commonly used to evaluate value estimation techniques. The lower the error metrics, the better.

The MAPE is a percentage relative error, as in (2), that expresses how much the absolute error between the real value ($y_i$) and the predicted value ($\hat{y}_i$) is greater than the real value, for a point $i$ in the time series. According to Yorucu [22], a prediction with MAPE percentage below 10 % is interpreted as highly accurate; forecast greater than 10 % and less than 20 % is interpreted as good; forecast greater than 20 % and less than 50 % is reasonable; and prediction greater than 50 % is considered inaccurate.

$$MAPE = \frac{1}{n}\sum_{i}^{n}\left|\frac{y_i - \hat{y}_i}{y_i}\right| \quad (2)$$

The sMAPE is a percentage measure of prediction errors according to (3) and indicates how much the observed error is greater than the sum of the modules of the real value ($y_i$) and the predicted value ($\hat{y}_i$), for the $N$ available points. Therefore, the closer to zero, the better the prediction.

$$sMAPE = \frac{100\%}{n}\sum_{t=1}^{n}\frac{|\hat{y}_t - y_t|}{|\hat{y}_t| + |y_t|} \quad (3)$$

MAE is simply the average of the absolute values of errors, i.e., the differences between the real value ($y_i$) and the predicted value ($\hat{y}_i$), according to (4).

$$MAE = \frac{1}{n}\sum_{i=1}^{n}|\hat{y}_i - y_i| \quad (4)$$

The proposed validation metrics are commonly used in prediction tasks. Thus, although the MAE is useful to verify the magnitude of the errors found along the predictions, the visualization together with MAPE and sMAPE is very important to understand the performance of the proposed method.

## IV. Results and Discussion

This section presents the results of the first experiments needed for parameter definition, case studies and the final result of the proposed method.

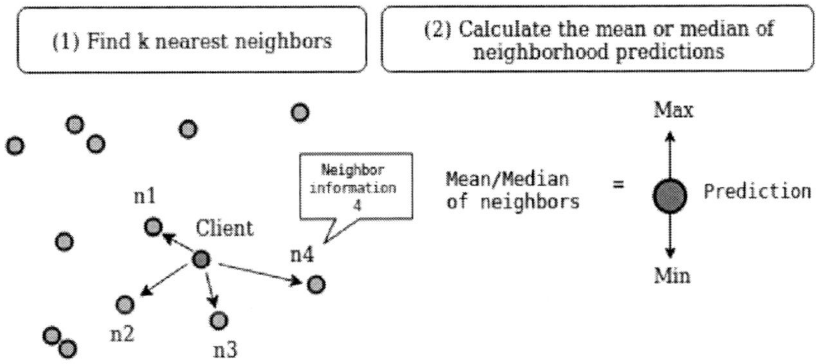


Figure 3. First scenario: a new customer with no history.

### A. *First experiments*

The first experiments were performed in the data set of the Residential class of a municipality X of Brazil. This dataset contains 107,738 customers. The objective of these experiments was to verify the influence of the calculation of distances on the results for the first scenario, where both consumption and interval predictions are generated only by the nearest neighbors. Thus, the Euclidean and Manhattan distances were tested. First, 10 % of these customers were randomly separated. We then simulated short series using only the customer's first four months and applied the consumption and interval estimation method described for the first scenario (III-E, using the predicted consumption median of the 10 nearest customers. Next, the result found for the Manhattan distance was an sMAPE of 35.4 % and the Euclidean distance obtained an sMAPE of 35.42 %. Despite the slight difference, the distance from Manhattan was chosen to be applied in real cases.

*1) Study of case:* To exemplify tests performed in the first experiments during the development of the methodology, specific cases to be analyzed were removed.

Fig. 4 presents a case considered highly accurate, according to Yorucu [22], with a 4.28 % sMAPE, an 8.20 % MAPE and a 6.14 MAE. In this case, the customer has, in the reference month analyzed, real consumption of 75 kWh and had a predicted consumption of 68.85 kWh. The neighborhood consumption range relative to the customer is close, resulting in a closer prediction of the real.

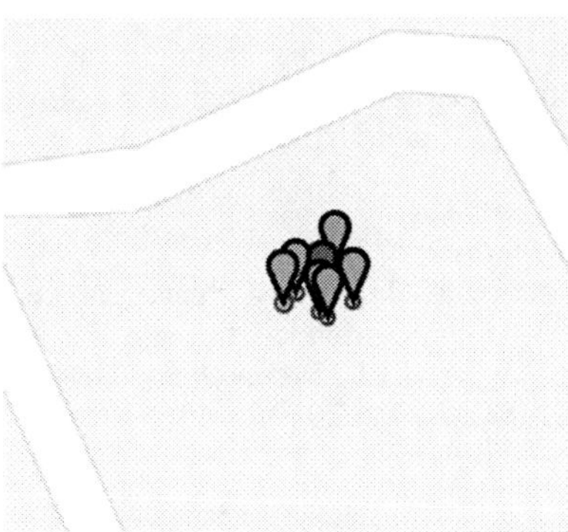

Figure 4. First case study: high accuracy of consumption prediction.

Fig. 5 presents a bad case, where the customer's consumption range is much lower than its neighborhood consumption range. In this case, the real customer consumption is 7 kWh, but the predicted was 358.21 kWh, with an sMAPE of 96.17 %, a MAPE of 351.21 % and an MAE of 5,017.29.

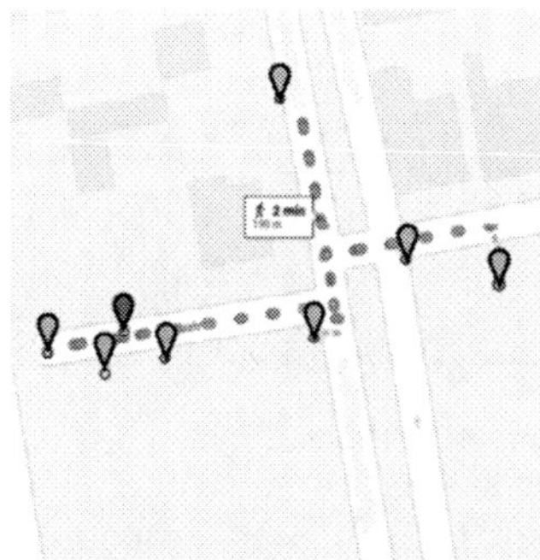

Figure 5. Second case study: low accuracy of consumption prediction.

### B. *Results of the proposed method*

This section describes the results of the experiment conducted with a short series of clients. In total, 86,874 customers remained after the preprocessing. In this experiment, the distance from Manhattan was used as a metric of the proximity of the points and ten neighbors closest to the new client were experimentally defined.

Table I shows the results by class of the methods chosen to predict neighborhood consumption, from the stage prior to the consumption forecast for new customers. The metric used to choose the methods was MAPE and each line represents a class with their respective values in each method used. From this same table, it is possible to verify that BAQR was the most used method among classes.

Table II presents the Quantity (QTY) of customers separated by class, which were processed by the proposed method and the results obtained in the consumption estimation for the short series. The residential class, with 79,871, contains the largest number of customers. While the lighting public class contains only 12. The best results were obtained with the sMAPE metric, as it is a symmetric metric, i.e., it limits the extreme effects, as well as avoiding null results. However,

TABLE I. RESULTS BY CLASS OF THE THREE BEST METHODS CHOSEN TO PREDICT NEIGHBORHOOD CONSUMPTION.

| | Methods | | |
|---|---|---|---|
| Classes | BAQR | STL | ARIMA |
| Residential | **36.20%** | 47.00% | 57.10% |
| Industrial | 41.80% | **32.90%** | 62.10% |
| Commercial | **32.60%** | 61.90% | 157.00% |
| Rural | **38.90%** | 48.70% | 203.70% |
| Government | 129.40% | 77.50% | **57.90%** |
| Lighting public | 29.50% | 32.90% | **25.10%** |
| Public service | **31.30%** | 337.80% | 229.70% |
| Self-Consumption | 14.40% | **10.00%** | 12.70% |

when analyzing the prediction results in terms of MAPE and MAE, these were not considered ideal but promising given the diversity of scenarios and consumption classes found in the dataset.

MAPE had the highest values where consumption patterns are highest. For example, in the class Self-consumption, with minimum consumption of 0, the average is 15,596 and a maximum of 82,740 kWh.

TABLE II. RESULTS OF VALIDATION METRICS BY CONSUMPTION CLASS.

| CLASSES | QTY. | sMAPE (%) | MAPE (%) | MAE | Within range (%) |
|---|---|---|---|---|---|
| Residential | 79,871 | 30.92 | 1,136.56 | 490.05 | 96.14% |
| Industrial | 76 | 26.57 | 133.29 | 379.28 | 94.74% |
| Commercial | 4,032 | 36.42 | 301.43 | 1,292.05 | 94.10% |
| Rural | 1,475 | 33.17 | 614.16 | 3,472.73 | 93.42% |
| Government | 577 | 25.15 | 91.11 | 443.54 | 93.41% |
| Lighting public | 12 | 22.71 | 45.12 | 667.67 | 91.67% |
| Public service | 98 | 41.57 | 200.91 | 1,548.66 | 83.67% |
| Self-Consumption | 733 | 22.69 | 447.65 | 86,477.52 | 89.50% |

Table III presents the percentage distribution of customers to sMAPE and MAPE percentage interval. The largest number of customers (37.39 %) were found with up to 10 % of the sMAPE metric. Analyzing sMAPE according to Yorucu's classification [22], 37 % of customers presented a highly accurate forecast, 16 % good, 22 % reasonable and only 23 % of customers presented an inaccurate forecast.

TABLE III. COSTUMERS DISTRIBUTION ACCORDING TO SMAPE AND MAPE METRIC VALUE RANGE.

| | Percentage of customers by metric | |
|---|---|---|
| Range | sMAPE (%) | MAPE (%) |
| [0 - 10%[ | 37.39 | 28.70 |
| [10% - 20%[ | 16.57 | 15.16 |
| [20% - 50%[ | 22.89 | 23.43 |
| [50% - inf[ | 23.15 | 32.70 |

During the results analysis, it was observed that close customers may have characteristics in common, such as purchasing power, consequently the same consumption pattern. However, there are areas where customer consumption does not have a common range, e.g., a new customer A, with a consumption range of 10 to 50 kWh, while its vicinity has a range of 100 to 300 kWh. Thus, it was found that new customers usually have low consumption in the first months or zero consumption, which was often incompatible with the consumption of their neighbors, who already have a stable consumption pattern. Therefore, this situation contributes to the increase of the prediction error.

Likewise, customers whose neighborhood may be located in a border region between two neighborhoods with different consumption characteristics, may experience an increase in the prediction error, since only the distance is used to determine that neighborhood.

The consumption prediction interval for new customers was not previously found by the company. Thus, the proposed method appears as an important tool for theses cases, due to the reasonable accuracy obtained. As a result, at least 83 % of the real cases were within the generated interval, in the worst case, as shown in the Table II, in Public service class.

This result is significantly relevant since this method, for this example, would prevent around 92 % (79,900) of these customers from going into the billing sector, avoiding the need for manual analysis of the recorded consumption for these customers. In the current practice of the company, all customers with short series end up going to this sector.

## V. Conclusion

In the present work, a method of prediction of power consumption for consumers with short or no consumption history was proposed. The method made use of machine learning techniques such as k-nearest neighbors, different distance measurements and various regressors such as SGD, STL, ARIMA, BAQR, LSTM, which are used to predict the energy consumption of consumers had their installation recently connected.

The proposed methodology used several approaches, such as Euclidean distance and Manhattan distance to find k-nearest neighbors; different regressors; and the median predicted neighborhood consumption. From the evaluated approaches, the distance from Manhattan showed a small advantage over Euclidean and among the regressors, the best estimates were made with STL, ARIMA, and BAQR.

From the above, it can be concluded that the neighborhood-based estimation method for power consumption is a promising method for new consumers, with no consumption history yet. In addition, it was possible to observe from the two selected case studies, that the proximity to neighbors results in a good result of prediction of consumption and its corresponding consumption interval, according to the first case study. The opposite, that is, when these neighbors are distant from each other, resulted in a low accuracy of prediction and their consumption range, as shown in Fig. 5, of the second case study. Therefore, the results presented contribute to reduce the volume of customers that need to be analyzed by the company.

Although promising, the method can be improved by utilizing other network-based serial data estimation techniques such as TCN [23] and N-BEATS [24].

## Acknowledgments

The authors would like to thank Equatorial Energy for the financial support provided through the National Electric Energy Agency (ANEEL) Research and Development Program (R&D), PD-00037-0036/2019.

## References

[1] A. Mosavi, M. Salimi, S. ardabili, T. Rabczuk, S. Shamshirband, and A. Varkonyi-Koczy, "State of the art of machine learning models in energy systems, a systematic review," Energies, vol. 12, 04 2019.

[2] J. Schneider, M. Dziubany, A. Schmeink, G. Dartmann, K.-U. Gollmer, and S. Naumann, "Chapter 8 - predicting energy consumption using machine learning," in Big Data Analytics for Cyber-Physical Systems, G. Dartmann, H. Song, and A. Schmeink, Eds. Elsevier, 2019, pp. 167 – 186, [accessed: 2019-10-10]. [Online]. Available: http://www.sciencedirect.com/science/article/pii/B9780128166376000087

[3] W. R. Tobler, "A computer movie simulating urban growth in the detroit region," Economic Geography, vol. 46, no. supl, 1970, pp. 234–240, [accessed: 2020-01-30]. [Online]. Available: https://www.tandfonline.com/doi/abs/10.2307/143141

[4] A. M. Alonso, F. J. Nogales, and C. Ruiz, "A single scalable lstm model for short-term forecasting of disaggregated electricity loads," 2019.

[5] K. M. Vu, The ARIMA and VARIMA time series: their modelings, Analyses and Applications. AuLac Technologies Inc., 2007.

[6] P. Goodwin, "Using naïve forecasts to assess limits to forecast accuracy and the quality of fit of forecasts to time series data (working paper)," 11 2014.

[7] A. Bâra and S. V. Oprea, "Electricity consumption and generation forecasting with artificial neural networks," in Advanced Applications for Artificial Neural Networks, A. El-Shahat, Ed. Rijeka: IntechOpen, 2018, ch. 7, [accessed: 2019-10-16]. [Online]. Available: https://doi.org/10.5772/intechopen.71239

[8] Z. Jiang, R. Lin, and F. Yang, "A hybrid machine learning model for electricity consumer categorization using smart meter data," Energies, vol. 11, no. 9, 2018, [accessed: 2019-10-15]. [Online]. Available: https://www.mdpi.com/1996-1073/11/9/2235

[9] F. Wahid and D. Kim, "A prediction approach for demand analysis of energy consumption using k-nearest neighbor in residential buildings," International Journal of Smart Home, vol. 10, 02 2016, pp. 97–108.

[10] A. T. Lora, J. R. Santos, J. R. Santos, J. L. M. Ramos, and A. G. Exposito, "Electricity market price forecasting: Neural networks versus weighted-distance k nearest neighbours," in Database and Expert Systems Applications, A. Hameurlain, R. Cicchetti, and R. Traunmüller, Eds. Berlin, Heidelberg: Springer Berlin Heidelberg, 2002, pp. 321–330.

[11] J. Poloczek, N. A. Treiber, and O. Kramer, "Knn regression as geo-imputation method for spatio-temporal wind data," in International Joint Conference SOCO'14-CISIS'14-ICEUTE'14, J. G. de la Puerta, I. G. Ferreira, P. G. Bringas, F. Klett, A. Abraham, A. C. de Carvalho, Á. Herrero, B. Baruque, H. Quintián, and E. Corchado, Eds. Cham: Springer International Publishing, 2014, pp. 185–193.

[12] M. Kim, S. Park, J. Lee, Y. Joo, and J. K. Choi, "Learning-based adaptive imputation methodwith knn algorithm for missing power data," Energies, vol. 10, no. 10, 2017, [accessed: 2019-10-16]. [Online]. Available: https://www.mdpi.com/1996-1073/10/10/1668

[13] N. E. E. Agency. Normative resolution no. 414/2010. [Online]. Available: www.aneel.gov.br/cedoc/ren2010414.pdf [retrieved: oct, 2019]

[14] ——. Normative resolution no. 800/2017. [Online]. Available: http://www2.aneel.gov.br/cedoc/ren2017800.pdf [retrieved: oct, 2019]

[15] R. B. Cleveland, "Stl : A seasonal-trend decomposition procedure based on loess," 1990.

[16] S. B. Taieb, R. Huser, R. J. Hyndman, and M. G. Genton, "Forecasting uncertainty in electricity smart meter data by boosting additive quantile regression," IEEE Transactions on Smart Grid, vol. 7, no. 5, 2016, pp. 2448–2455.

[17] D. M. Nelson, A. C. Pereira, and R. A. de Oliveira, "Stock market's price movement prediction with lstm neural networks," in 2017 International Joint Conference on Neural Networks (IJCNN). IEEE, 2017, pp. 1419–1426.

[18] S. Ruder, "An overview of gradient descent optimization algorithms," arXiv preprint arXiv:1609.04747, 2016.

[19] D. Kraus and C. Czado, "D-vine copula based quantile regression," Computational Statistics Data Analysis, vol. 110, 2017, pp. 1 – 18, [accessed: 2019-10-15]. [Online]. Available: http://www.sciencedirect.com/science/article/pii/S0167947316303073

[20] R. J. Hyndman, G. Athanasopoulos, and OTexts.com, Forecasting : principles and practice / Rob J Hyndman and George Athanasopoulos, print edition. ed. OTexts.com [Heathmont?, Victoria], 2014 2014.

[21] B. Etienne, "Time series in python - exponential smoothing and arima processes," Mar 2019, [accessed: 2019-10-11]. [Online]. Available: https://towardsdatascience.com/time-series-in-python-exponential-smoothing-and-arima-processes-2c67f2a52788

[22] V. Yorucu, "The analysis of forecasting performance by using time series data for two mediterranean islands," Review of Social, Economic & Business Studies, vol. 2, 2003, pp. 175–196.

[23] F. Karim, S. Majumdar, H. Darabi, and S. Chen, "Lstm fully convolutional networks for time series classification," IEEE Access, vol. 6, 2017, pp. 1662–1669.

[24] B. N. Oreshkin, D. Carpov, N. Chapados, and Y. Bengio, "N-beats: Neural basis expansion analysis for interpretable time series forecasting," arXiv preprint arXiv:1905.10437, 2019.

[25] M. Laurinec, P. & Lucká, "Clustering-based forecasting method for individual consumers electricity load using time series representations," Open Computer Science, no. 8, 2018, pp. 38–50.

# Harmonized Multiresolution Geodata Cube for Efficient Raster Data Analysis and Visualization

Lassi Lehto, Jaakko Kähkönen, Juha Oksanen and Tapani Sarjakoski
Geoinformatics and Cartography
Finnish Geospatial Research Institute (FGI)
National Land Survey of Finland
Masala, Finland
email: lassi.lehto@nls.fi, jaakko.kahkonen@nls.fi, juha.oksanen@nls.fi, tapani.sarjakoski@nls.fi

*Abstract*— **The Data Cube concept provides a useful metaphor for management of raster geodata resources in the cloud. An initiative, called GeoCubes Finland, has been launched with the aim to facilitate access to geospatial raster data for academic research. The work is carried out in the context of a major research infrastructure development program in Finland. In the ingestion process, data sets are pre-processed into a harmonized, multiresolution cloud-based repository and brought into a common georeferencing frame, resolution levels, encoding format and tiling scheme. A custom Application Programming Interface (API) has been developed for flexible query, download and analysis of the repository's content layers. Cloud-optimized access to pre-stored resolution levels supports efficient interactive visual exploration of analysis results computed on-the-fly.**

***Keywords-raster data; multi-resolution; harmonisation; cloud service; visualization.***

## I. INTRODUCTION

Geospatial data sets are growing rapidly in number and volume. In particular, the volume of raster data sets is becoming difficult to manage. The resolution of image sensors has steadily improved and new imaging technologies have been taken into use. In addition, integrating geospatial raster data sets for analysis has become a tedious task, as the data sets typically differ in critical parameters like origin, resolution, coordinate reference system, encoding, format, etc. It important to develop mechanisms that facilitate access to relevant data resources.

The data cube concept has emerged as a solution for organizing a repository of harmonized geospatial raster datasets [1] [2]. In computer technology, a data cube is a multi-dimensional array of values [3]. Those values represent certain fact that can be organized along various aspects. For instance, results of a vote can be considered along political party, voting districts, age or gender of candidates, year of election, etc. These aspects correspond to the axes of the multi-dimensional array and the values live in cells inside this array.

In the geospatial domain, the data cube approach was first used in the Earth Observation (EO) community for organizing vast amounts of satellite images [4]. In this application area time is a very important dimension, as satellite imagery are typically available in extensive time series. The four most typical data cube dimensions in an EO application thus are the two geospatial coordinate axes, time and the imagery type.

In the case of a geospatial data cube, the predominant axes naturally are latitude, longitude and the possible height. A voxel-based approach for organizing truly 3D geodata is also possible [5]. The cell value represents the fact about the physical environment that the data set happens to describe. Thus, content theme can be seen as a predominant axis of a geospatial data cube. The most important benefit of organizing data sets as a data cube repository is immediate availability of the contents for integrated analysis. The approach thus aims at fulfilling the goals of the concept Analysis Ready Data (ARD) [6].

The most important positive aspects that a multidimensional data cube approach provides for geospatial application can be listed as

- A data cube forms an integrated data repository, where harmonized content layers are ready for use without troublesome data preparation steps
- Ingested datasets are aligned on pixel level and thus can be readily integrated for visualization and analysis
- A data cube can be accessed along any of its dimensions, allowing for new kinds of knowledge retrieval and spatial analysis processes
- A joint data cube repository enables data providers to deliver content through a new, easy-to-use channel, thus expanding their user base
- The simplicity of raster data processing can be more effectively and widely exploited
- By organizing a data cube as a cloud-based service with Web-friendly APIs, the potential use scenarios can be further widened

The importance of the data cube concept for geospatial application domain is also demonstrated by the fact that the Open Geospatial Consortium has initiated standardization work on the subject [7]. Another significant development is the Open Data Cube (ODC) initiative. ODC is an open source software library for organizing vast amounts of satellite imagery according to principles of data cube [8]. The software includes components for cataloguing and indexing EO resources and for ingesting those resources into a harmonized and optimized storage format for easy data retrieval. Furthermore, the ODC library provides tools for accessing and downloading content and for performing analysis operations on it. The ODC community has also developed tools for visual exploration and statistical analysis of the data cube contents. Concrete national and regional data cube implementations based on ODC include the Australian data cube called Digital Earth Australia [9], the Swiss Data Cube [10], Columbian Cube, Vietnam Open Data Cube, and the Africa Regional Data Cube covering Kenya, Senegal, Sierra Leone, Ghana and Tanzania [11].

ODC can also be applied to data resources other than EO data. The examples mentioned in the ODC documentation include gridded data sets like elevation models, geophysical grids and other interpolated surfaces. However, most of the activities around geospatial data cubes focus on satellite image processing, in general, and on their time series applications, in particular [12].

An initiative has been launched in Finland to build a multi-resolution, cloud service-based geodata cube, called GeoCubes Finland, containing some of the most important national geodatasets [13]. The contents of the GeoCubes include data layers like Digital Elevation Model (DEM), administrative areas, land use, surface deposits and various attributes of the national forest inventory. The content layers are provided by the governmental agencies that collect and continuously maintain them. GeoCubes Finland thus differs from other geodata-related data cube implementations by not focusing on satellite imagery, but rather on other traditional geodata sets. The GeoCubes development is being carried out in the context of a large research infrastructure development program in Finland, called Open Geospatial Information Infrastructure for Research (oGIIR) [14].

In Section II, the basic principles of the GeoCubes Finland are presented. Section III discusses the custom content access API of the GeoCubes Finland data repository in more detail. In Sections IV and V, some visualization related considerations and example applications are discussed. Section VI concludes the paper.

## II. GeoCubes Finland

### A. General

A particular feature that makes GeoCubes Finland different from other geospatial data cube implementations is its multi-resolution nature. A set of fixed resolution levels (1, 2, 5, 10, 20, 50, 100, 200, 500 and 1000 m) have been selected to store the contained data sets. The resolutions are selected to facilitate processing of analysis operations, typically run on round resolution values, and to enable easy integration with statistical and other auxiliary data sets. Opposite to resolution levels typical for web mapping schemas, such as the powers of two as in Google Maps, in GeoCubes Finland the resolution levels are selected to serve human user. The actually available set of resolution levels depend on the original accuracy of the source data set. Thus, the finest resolution of most of the GeoCubes Finland's content layers is 10 m. Only some national data sets can be reasonably be represented in 1 m resolution. These include the most accurate digital elevation model and the administrative division data sets.

Resolution levels could possibly be seen as one dimension of a multi-dimensional geodata cube, but each resolution level has an individual range for the cube's coordinate-related axes. Therefore, the resolution levels of the cube must actually be seen as a set of separate cube instances. Hence, the plural form of the cube's name: GeoCubes Finland.

In addition to industry-standard access interface to coverage data, the Web Coverage Service (WCS) [15], the GeoCubes contents can be accessed via direct file URLs and through the custom-built GeoCubes API.

### B. Technical Implementation of the Repository

GeoCubes Finland's data storage is implemented as a set of Cloud Optimized GeoTIFF (COG) files [16] [17]. These files are stored on a cloud service platform, organized in directories by data provider, data set and the edition of data set. The area of the country is divided into tessellation of sixty 100 km * 100 km sized blocks, each maintained as a separate GeoTIFF file to ease processing tasks. The tessellation also facilitates parallelization of various processing steps and works as a rudimentary spatial index. All the files can always be accessed in a straightforward manner via http (HyperText Transfer Protocol) by their URLs (Uniform Resource Locator).

A GeoTIFF file arranged according to COG specification contains overviews and has its raster content organized as tiles. The overviews are ordered in sections from the coarsest to the most detailed, followed by the full resolution raster. The metadata describing the offset of each section has to be at the beginning of the file. With this file structure, a calling COG-aware client can make use of the http 'GET Range' query, which enables an indicated subsection of a file to be requested. By first requesting the metadata portion, the client can then select an appropriate range of bytes to only download the needed geospatial area of an appropriate resolution, i.e. overview level. This mechanism significantly improves the efficiency of raster data retrieval for cloud-service based applications.

The multiple resolutions are implemented in two ways, both as internal GeoTIFF overview layers and as individual external resolution-specific GeoTIFF files. Internal overviews facilitate easy content delivery, as all resolutions are contained inside as single file. Separate resolution-specific GeoTIFF files in turn enable better control of resolutions for computations run in external applications. The overview functionality, like most other computing tasks in GeoCubes Finland, is implemented using Geospatial Data Abstraction Library (GDAL) [18].

To collect the individual files as a complete view covering the whole country, the so-called Virtual Format mechanism (VRT) of GDAL is used [19]. In this approach, an XML-formatted text file is used to refer to the set of image files that constitute the integrated view. As such, a single VRT file can for instance refer to all external GeoTIFF overview files on certain resolution level, thus facilitating analysis and visualization of larger areas. In the same way, a VRT file can refer to various resolution-specific files in a certain tessellation block, providing a light-weight representation of the multi-resolution data set of that block. VRT files can be easily transported across network connections. Because the file references are stored as absolute addresses, the VRT file can be opened in a third-party application. This approach limits downloading of raster content on the spatial area and on the resolution level the user actually needs.

## III. CONTENT ACCESS API

An API has been developed for exploring, downloading and analyzing the GeoCubes Finland contents [20]. The API is designed according to the principles of RESTful Web API [21], in which the request is defined by the path components of the query URL. The API provides requests for querying the basic metadata of the repository, querying cell values on a given location, downloading content by bounding box, polygon or administrative unit and analyzing the content in terms of cell value distribution, change detection, etc. The analysis operations are run on the server platform, without downloading any content to the client side.

In all content queries, the query can be run on a desired resolution level and year. The multi-resolution structure of the GeoCubes repository gives to the user a flexible choice on speed vs. accuracy of the operation. For instance, an analysis procedure under development can be first tested on a coarse resolution level, and the real, time-consuming run on the finest resolution be carried out only when seemed appropriate. In case of analysis based on visual exploration, the resolution level, on which the analysis is run, can always be matched with the zoom level of the visualization. This way the analysis can be run in roughly constant time and the interactivity level of the application can be kept stable.

The general scheme of the access API can be described as follows:

```
what to do /
   on which resolution level /
      with which content layer /
         where /
            when /
               how
```

As an example, the query for the cell value on a given location becomes as:

```
legend/500/mvmi-paatyyppi/340500,6695000/2009
```

where mvmi-paatyyppi (in Finnish) is one of the themes (forest inventory data) of GeoGubes.

Another query would request download of a content layer inside the given list of municipalities ('kuntajako' in Finnish), from the given resolution level and year:

```
clip/100/corine/kuntajako:734,761,834,433,224
,444,927/2000
```

The resulting data set, requested using the GeoCubes Web client, is shown in Figure 1.

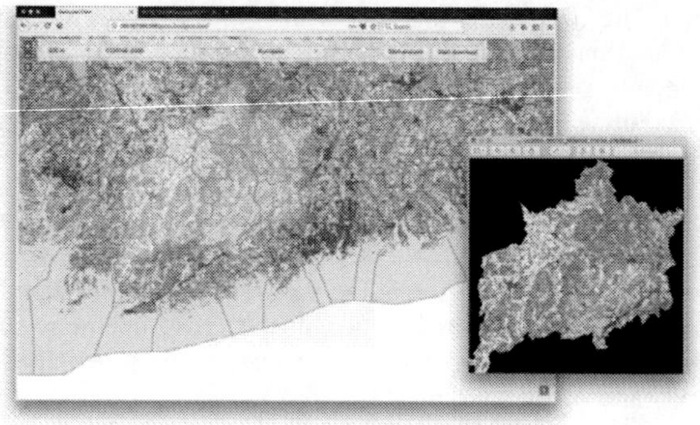

Figure 1. A CORINE data set downloaded via GeoCubes API using administrative units' boundaries as selection area.

Downloading of a DEM from two blocks in VRT format with multiresolution content would go as:

```
clip/20/km10/blocks:300000,6900000,
300000,6800000/2018/vrt/mr
```

An example of an analysis would be change detection between CORINE versions 2000 and 2012 in land use type fields ('pellot' in Finnish) inside the given bounding box:

```
analyse:changedetect/10/corine:Pellot
/bbox:225700,6660000,494300,6740000/2000,2012
```

The result of the above analysis is an image depicting the changes in red (removals) and green (additions).

Another analysis would determine the distribution of cell values inside the two indicated counties ('maakuntajako' in Finnish):

```
analyse:distribution/500/mvmi-
maaluokka/maakuntajako:04,06/2015
```

The result is a list of existing cell values together with their frequencies. The analysis results visualized in the GeoCubes Web client using the D3.js library [22] is shown in Figure 2.

Figure 2. GeoCubes analysis results showing distribution of data values inside the selected area.

As an auxiliary data source for analysis purposes, the Finnish administrative areas in four different levels are also available in vector form in the GeoCubes repository.

The GeoCubes API has been implemented as a Web service using Django Web framework [23], together with Python-based service side scripts making an extensive use of the Python API of GDAL. In addition to the custom API, the most relevant service interfaces standardized by the Open Geospatial Consortium are also supported. The main components of the GeoCubes Finland platform are shown in Figure 3.

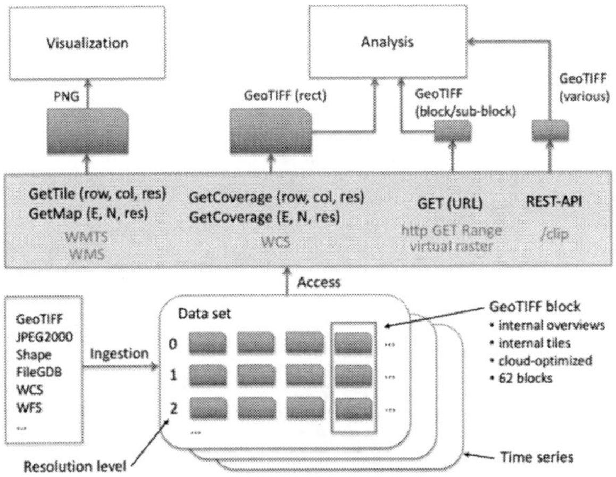


Figure 3. The main system components of the GeoCubes Finland platform.

## IV. VISUALIZATION CONSIDERATIONS

For basic visualization of the content layers, a Web Map Tile Service (WMTS) is available on the GeoCubes platform. The service makes use of a Web Map Service (WMS) that uses a VRT file combining the original GeoTIFF files as its source. This arrangement enables both ad hoc visualizations through WMS and cache-based static image delivery via WMTS. In the case of the WMTS service, the pre-rendered tiles are available on the fixed resolution levels of the GeoCubes. For best visual presentation, the client should apply scale levels that correspond to those resolutions. The WMS implementation of GeoCubes platform is based on MapServer and the WMTS service on MapProxy.

Visualization of administrative areas poses a specific challenge. Because areas are represented as raster data to facilitate analysis with other content layers, boundary lines cannot be used to separate an area from the neighboring areas. The only way to reliably display the administrative division of the country in a legible manner is to ensure that neighboring areas are always presented with clearly distinguishable colors. The Finnish lowest level administrative division has more than 300 municipalities. Thus, to achieve an appropriate data set with cell values corresponding to the real municipality codes, a raster with 16-bit cell values was created. However, for Web browser-based visualization, a 256-colour PNG image is required. A custom software module was developed that creates random color components for a 256-colour RGB color table and applies the colors to municipalities so that the three color components of a municipality always differ more than 30 units from all of the neighboring areas' color components. The result is a bright and colorful municipality map, in which all areas can be reliably distinguished (Figure 4).

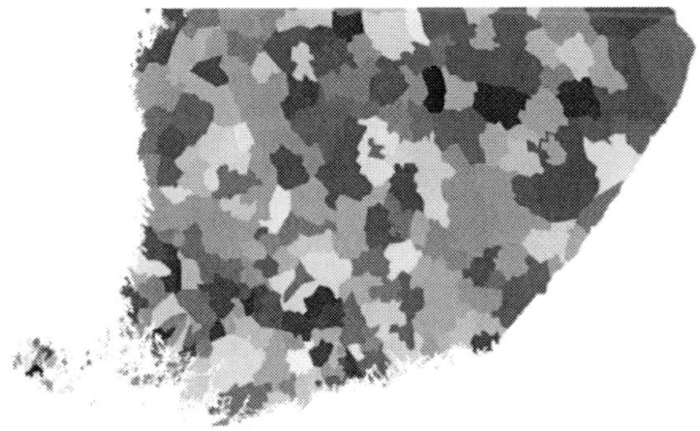

Figure 4. A municipality map with well distinguishable colors.

Another example of enhanced visualization tested in GeoCubes Finland is the locally stretched DEM visualization. The terrain of the country is mostly flat, higher elevations existing only in the northern Lapland. A consistent, stable country-wide DEM visualization thus depicts flat areas with very limited color scale, making it impossible to recognize subtle height differences. A dynamic visualization module was developed for GeoCubes that stretches every individual view requested by the client to full color scale. As the GeoCubes data repository has multiple resolution levels, the visualization processing can be performed efficiently in constant time, independently of the presentation scale of the client. The two images in Figure 5 depict the difference between fixed and dynamic DEM visualization on a particularly flat area in the Finnish Ostrobothnia.

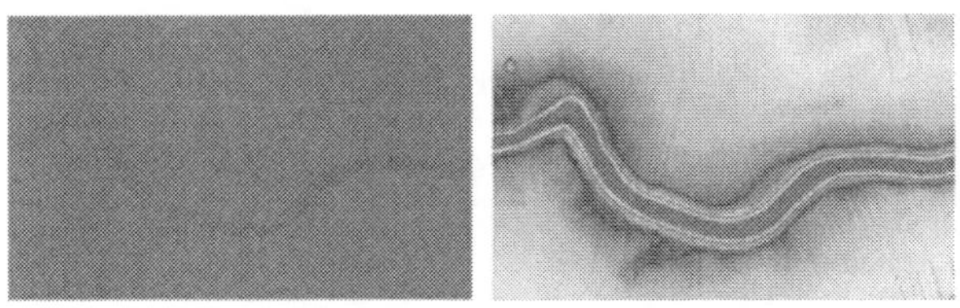

Figure 5. A static DEM visualization compared with a dynamic, locally stretched graphic scale.

A 3D example of visualization of GeoCubes' contents is depicted in Figure 6. Here, DEM data is extracted from the GeoCubes API around possible crater locations, where a meteorite impact is deemed to be behind the peculiar rounded land form. The multiresolution contents of the GeoCubes repository supports effective 3D visualization of crater areas of various sizes.

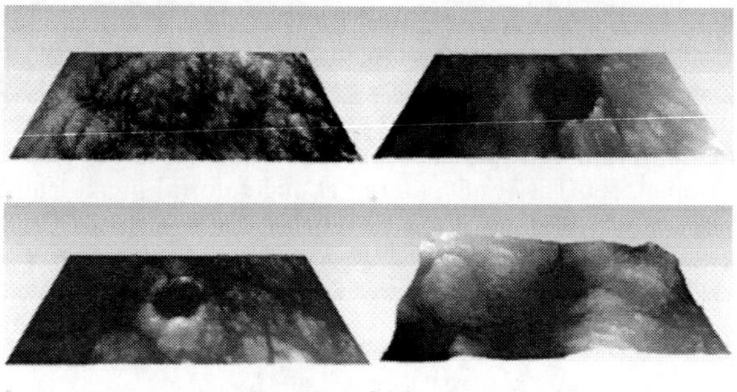

Figure 6. Possible meteorite impact craters visualized, based on GeoCubes DEM. Big scale differences in visualizations are supported by GeoCubes' multiresolution contents.

## V. APPLICATION EXAMPLES

The analysis processes can be effectively run on different resolution levels and their results are readily available for visual exploration over the whole range of scales supported by the geodatacube's resolutions. The GeoCubes API has been designed to support access, querying and analysis of the geodatacube's contents on the resolution level appropriate for the actual use situation.

### *A. Field Inspection*

As an example of using GeoCubes API, a field inspection of repository cell values has been tested. In this approach, the inspector moves around on the terrain and constantly receives values of the GeoCubes layer of interest into his cell phone. This way, the inspector can compare the environment around him with the stored category values, for instance, he can determine the correctness of forest type information – and do that on all available resolution levels of the repository.

In the developed pilot implementation, the position of the inspector is recorded with an application called Owntracks. The location is reported back to a server (Eclipse Mosquitto) [24] implementing the ISO-standardized Message Queuing Telemetry Transport (MQTT) -protocol. The user is presented with a map-based application (Leaflet with Realtime extension). When the map application gets notice from the MQTT server that the user has moved to a new location, it then send a request to the 'legend' operation of the GeoCubes API together with the location information. As a result, it will get cell values from the requested layer on all available resolution levels and can compare them with the reality around him. The pilot's architecture is presented in Figure 7.

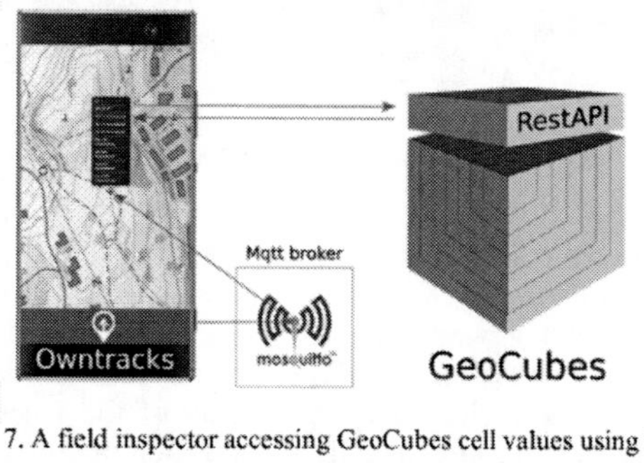


Figure 7. A field inspector accessing GeoCubes cell values using Mosquitto server.

### *B. Route Finding*

In another analysis example, the GeoCubes DEM layer is used for route finding for a forest vehicle that has certain limits for movement in steep slopes. DEM data for study area was acquired by accessing the GeoCubes API's 'clip' operation from within the QGIS application [25]. Then, analysis functions available in QGIS were used to compute slope values for the area and these were used as the cost surface for route finding. Areas too steep for the vehicle were excluded from the computation. The case study demonstrates the use of the GeoCubes' content for analysis using tools outside the GeoCubes platform. The same analysis could also be run in different resolution levels, depending on the needs of the application. The resulting route alternatives are depicted in Figure 8 using three.js, a 3D JavaScript library [26].

Figure 8. Route finding in QGIS with DEM from GeoCubes using slope as cost surface. White areas denote terrain that has been classified to be too steep and is excluded. The red and yellow lines denote found alternative routes.

## VI. CONCLUSIONS

One of the major obstacles for wider use of geospatial raster data sets in different research and analysis scenarios is the work required for pre-processing the available data sets. This often involves coordinate reference system transformations, resampling procedures, coding system translations, integration of map sheet-based data files, etc. To facilitate the introduction of geospatial data sets into research processes in a multidisciplinary setting, a harmonized, easy-to-access data storage would be really beneficial. In the GeoCubes Finland initiative, this kind of approach has been taken.

A representative set of geospatial data sets with national coverage has been ingested into the cloud service-based data repository. The pre-processing phase involves operations like rasterizing vector-formatted source data sets, resampling of source data into the set of standardized GeoCubes resolution levels, harmonization of value coding systems between different data set editions, encoding of the resulting raster content into the common cloud-optimized image format and storing it to the cloud repository, both as binary images files and as textual VRT representations.

By storing the raster data sets in multiple resolution levels, the GeoCubes repository can support certain use cases very efficiently. These include interactive visual exploration of on-the-fly analysis results and testing of an analysis procedure on coarse resolution levels before launching an accurate analysis on detailed levels. The result layer of an analysis can be configured as a new GeoCubes content layer and thus be run dynamically by the calling visualization client. The analysis procedure will always select the resolution level closest to the visual scale, thus enabling nearly constant processing times to be maintained.

Further work on the GeoCubes Finland platform will focus on adding new layers to the repository's data collection, developing modular analysis functions for server-side processing, and enhancing the GeoCubes Web client. A QGIS plugin module for accessing the GeoCubes API is also going to be developed. More user testing will be carried out to gather feedback and guidance for further development of the platform.

## ACKNOWLEDGMENT

We made use of computing resources provided by the Open Geospatial Information Infrastructure for Research (oGIIR, urn:nbn:fi:research-infras-2016072513) funded by the Academy of Finland, and CSC – The IT Center for Science Ltd.

## REFERENCES

[1] P. Baumann, The Datacube Manifesto, 2017. http://earthserver.eu/ sites/default/files/upload_by_users/The-Datacube-Manifesto.pdf [retrieved: Jan 2020].

[2] L. Lehto, J. Kähkönen, J. Oksanen, and T. Sarjakoski. GeoCubes Finland - A Unified Approach for Managing Multi-resolution Raster Geodata in a National Geospatial Research Infrastructure. In: *GEOProcessing 2018, the Tenth International Conference on Advanced Geographic Information Systems, Applications and Services*, March 25-29, 2018, Rome, Italy. ISBN: 978-1-61208-617-0, pp. 18-22.

[3] Wikipedia, Data cube. https://en.wikipedia.org/wiki/ Data_cube, 2020 [retrieved: Jan 2020].

[4] A. Lewis et al., "Rapid, high-resolution detection of environmental change over continental scales from satellite data – the Earth Observation Data Cube". *International Journal of Digital Earth*, vol. 9 , Iss. 1, 2016, pp. 106-111.

[5] U. Pyysalo and T. Sarjakoski, "Voxel approach to landscape modelling". *The International Archives of the Photogrammetry and Remote Sensing*, July 2–11, 2008, Beijing, China, XXXVII(B4/1), pp. 563–568.

[6] G. Giuliani et al. Building an Earth Observations Data Cube: lessons learned from the Swiss Data Cube (SDC) on generating Analysis Ready Data (ARD), Big Earth Data, 2017, 1:1-2, pp. 100-117, DOI: 10.1080/20964471.2017.1398903.

[7] OGC, Datacube Domain Working Group Charter, https://external.opengeospatial.org/twiki_public/pub/CoveragesDWG/Datacubes/17-071_Datacube-DWG_Charter.pdf, 2017 [retrieved: Jan 2020].

[8] ODC, Open Data Cube Home Page, https://www.opendatacube.org, 2019 [retrieved: Jan 2020].

[9] A. Lewis et al., "The Australian Geoscience Data Cube – Foundations and lessons learned". *Remote Sensing of Environment*, 2017, ISSN 0034-4257, https://doi.org/10.1016/j.rse.2017.03.015, pp. 276-292.

[10] Swiss Data Cube (SDC) Home Page, https://www.swissdatacube.org, 2017 [retrieved: Jan 2020].

[11] Africa Regional Data Cube Home Page, http://www.data4sdgs.org/index.php/initiatives/africa-regional-data-cube, 2016 [retrieved: Jan 2020].

[12] A. Lewis et al., 2016. Rapid, high-resolution detection of environmental change over continental scales from satellite data - the Earth Observation Data Cube. Int. J. Digital Earth 9 (1), pp. 106–111.

[13] L. Lehto, J. Kähkönen, J. Oksanen, and T. Sarjakoski. Supporting Wide User-Base in Raster Analysis - GeoCubes Finland. *The International Archives of the Photogrammetry, Remote Sensing and Spatial Information Sciences*, XLII-4, pp. 329-334. https://doi.org/10.5194/isprs-archives-XLII-4-329-2018.

[14] oGIIR, Open Geospatial Information Infrastructure for Research Home Page, http://ogiir.fi, 2020 [retrieved: Jan 2020].

[15] OGC, Web Coverage Service. http://www.opengeospatial.org /standards/wcs [retrieved: Jan 2020].

[16] GeoTIFF, GeoTIFF home page, http://trac.osgeo.org/geotiff/, 2019 [retrieved: Jan 2020].

[17] Cloud Optimized GeoTIFF (COG) Home Page, https://www.cogeo.org, 2019 [retrieved: Jan 2020].

[18] GDAL, Geospatial Data Abstraction Library Home Page, http://gdal.org, 2020 [retrieved: Jan 2020].

[19] GDAL, Virtual File Tutorial, https://www.gdal.org/gdal_vrttut.html, 2019 [retrieved: Jan 2020].

[20] L. Lehto, J. Kähkönen, J. Oksanen, and T. Sarjakoski, 2019. Flexible Access to a Harmonised Multi-resolution Raster Geodata Storage in the Cloud. *GEOProcessing 2019, the Eleventh International Conference on Advanced Geographic Information Systems, Applications and Services*, Feb 24-28, 2019, Athens, Greece. ISBN: 978-1-61208-687-3, pp. 26-28.

[21] RESTful API Tutorial, https://searchmicroservices.techtarget.com/definition/RESTful-API, 2020 [retrieved: Jan 2020].

[22] M. Bostock, 2018. Data-Driven Documents, D3.js Home Page. https://d3js.org [retrieved: Jan 2020].

[23] Django Software Foundation, 2018. Django Home Page. https://www.djangoproject.com [retrieved: Jan 2020].

[24] R. A. Light, "Mosquitto: server and client implementation of the MQTT protocol," The Journal of Open Source Software, vol. 2, no. 13, May 2017, DOI: 10.21105/joss.00265.

[25] QGIS Home Page, https://qgis.org/en/site/, 2020 [retrieved: Jan 2020].

[26] three.js Home Page, https://threejs.org, 2020 [retrieved: Jan 2020].

# Using Natural Language Processing for Extracting GeoSpatial Urban Issues Complaints from TV News

Rich Elton Carvalho Ramalho, Anderson Almeida Firmino, Cláudio de Souza Baptista, Ana Gabrielle Ramos Falcão and Maxwell Guimarães de Oliveira

Information System Laboratory
Computer Science Department
Federal University of Campina Grande (UFCG)
Campina Grande - PB, Brazil
Email: rich.ramalho@ccc.ufcg.edu.br, andersonalmeida@copin.ufcg.edu.br, anagabriellee@gmail.com, baptista@computacao.ufcg.edu.br, maxwell@computacao.ufcg.edu.br

Fábio Gomes de Andrade

Federal Institute of Paraíba (IFPB)
Cajazeiras - PB, Brazil
Email: fabio@ifpb.edu.br

***Abstract*—Citizens as sensors enable the engagement of society through technology to complain on urban issues. Despite the fact that some geosocial networks have been developed in recent years to enable citizens to report many types of urban problems, it is possible to notice that the engagement of the users of these networks usually decreases in time. Hence, many relevant issues are not identified or published, which reduces the effectiveness of these networks. Aiming to overcome this limitation, this paper proposes an approach in which urban issues are automatically detected from a TV news program. The proposed solution uses geoparsing and Natural Language Processing (NLP) techniques to geocode and classify the identified complaints and publishes the results in Crowd4City, a geosocial Network that deals specifically with urban issues. Finally, our method was evaluated using data of a real news TV program in Brazil. Our results indicate 59.8% of success on extracting text and location from the video news.**

***Keywords–Geosocial network; NLP; Urban Issues; Crowdsourcing.***

## I. Introduction

The high concentration of population in urban areas has imposed to local authorities several challenges to address issues concerning mobility, security, infrastructure, education, health, etc. These are what we call urban issues. One important challenge for these authorities consists of identifying the problems that have been faced by citizens.

Aiming to solve this limitation, in the context of Smart Cities, some authors developed geosocial networks that deal specifically with urban issues. These networks enable the use of context aware services to locate users and their complaints.

In the context of Smart Cities, geosocial networks enable the use of context aware services to locate users and their complaints on urban issues. Several tools, such as Crowd4City [1], Wegov [2] and FixMyStreet [3] have been proposed in order to provide the citizen an opportunity to complain on urban issues. However, people's motivation in using such geosocial networks decrease in time. Hence, to ensure a high engagement of society, different approaches to gather information are required.

Several local TV stations in Brazil portray urban issues reported by the community. An example is the 'Calendar' board in a daily open TV channel news program in the State of Paraíba, Brazil. That news broadcast exhibits several urban issues faced by the main cities from that particular state. Hence, it is important to gather those urban issues and input them into geosocial networks in order to improve citizenship and increase awareness. The audio descriptions in the news channel need to be converted into text, then geoparsing tools from Geographic Information Systems (GIS) and Natural Language Processing (NLP) techniques need to be used to automatically extract the correct location of the respective urban issues.

In this paper, we propose a framework to extract audio files from TV news, convert them into text documents, then extract location using a gazetteer and urban issues from text using NLP techniques in order to feed the Crowd4City geosocial network. It is important to mention that the news are up-to-date and extracted from a real context. Our main contribution consists in the integration of GIS and NLP.

The remainder of this paper is structured as follows. Section II discusses related work. Section III presents an overview of the Crowd4City geosocial network. Section IV focuses on our proposed method for extracting and structuring urban issues reported in TV news. Section V presents a case study and discusses the results. Finally, Section VI concludes the paper and points out further research to be undertaken.

## II. Related Work

NLP has been broadly used in several application domains including: machine translation, speech recognition, chatbots/question answering, text summarization, text classification, text generation, sentiment analysis, recommendation systems and information retrieval. Britz et al. [4] discuss machine translation using a seq2seq model. Reddy et al. [5] present a question answering approach. Schwenk et al. [6] focus on text classification. Radford et al. [7] propose a language model using unsupervised learners. NLP is difficult to accomplish as text differs from language to language.

Upon developing our proposed approach, we first performed an extensive study on the already existing models within scenarios similar to ours. Given that our method is based on NLP and geoparsing, we discovered some useful corpus. Oliveira et al. [8], for instance, contributed with the creation of a gold-standard corpus of urban issues related tweets in the English language, including geographical information. Such information can be very useful for improving geoparsers and for developing classifiers for the detection of urban issues. Focusing on our TV news domain, Camelin et al. [9] composed a corpus of different TV Broadcast News from French channels and online press articles, which were manually annotated in order to obtain topic segmentation annotations and linking annotations between topic segments and press articles. They made the FrNewsLink available online for anyone who wishes to use it in their studies. Although both corpora are based on different languages than the one used in our study, they proved to be useful in such domains.

Aiming at obtaining information from the TV news videos, Kannao and Guha [10] focused their study on extracting text from the overlay banner presented in such broadcasts. Such text usually contains brief descriptions of news events and, since they may be in various formats. Therefore, they proposed a contrast enhancement preprocessing stage and a parameter free edge density based scheme for better text band and text extraction. They also performed experiments using Tesseract Optical Character Recognition (OCR) for overlay text recognition trained using Web news articles. The authors confirmed the validity of their approaches using three Indian English television shows and obtained significant results. However, their domain is limited, since only the overlay bands' contents are analyzed.

Similarly, Pala et al. [11] developed a system for the transcription, keyword spotting and alerting, archival and retrieval for broadcasted Telugu TV news. Their main goal was to aid viewers in easily detecting where and when topics of their interest were being presented on TV news in real time and they were also hoping to assist anyone (including editorial teams at TV studios) in discovering videos of TV news reports about specific topics, defined by the user with keywords. Their system was the first that enabled the simultaneous execution of the broadcasted audio (speech), video and transcription of the audio in real time with the Indian Language, with keyword spotting and user alerts. Although it can detect topics of interest with the keyword, the system does not have the ability to extract the theme or domain being discussed in the video.

Bansal and Chakraborty [12] proposed an approach for content based video retrieval by combining several state-of-the-art learning and video/sentence representation techniques given a natural language query. They aimed at overcoming the robustness and efficiency problems found in the existing solutions using deep learning based approaches, combining multiple learning models. Their results show they were able to capture the videos' and sentences' semantics when compared to other already existing approaches, however the authors lack retrieving any geographic information.

Dong et al. [13] focused on developing a method for subject words extraction of urban complaint data posted on the Internet. Their approach consisted on the segmentation of the complaint information, extraction and filtering of candidate subject words, and was validated using 8289 complaints posted on a Beijing website. The proposed method showed that better results can be obtained than the Term Frequency–Inverse Document Frequency (TF-IDF) and TextRank methods in the context of written informal content made by Internet users. Nonetheless, such approach would need to be validated in other scenarios.

Mocanu et al. [14] proposed a method for such extraction by using temporal segmentation of the multimedia information, allowing it to be indexed and thus be more easily found by the users interested in specific topics. Their approach was based on anchor person identification, where the TV news program presenter would be featured on the video. They performed a few tests with a limited database of French TV programs and obtained good results, however their topic detection is not very robust, since it is based only on the video subtitles.

Zlitni et al. [15] addressed the problem of automatic topic segmentation in order to analyze the structure and automatically index digital TV streams, using operational and contextual characteristics of TV channel production rules as prior knowledge. They used a two-level segmentation approach, where initially the program was identified in a TV stream and then the segmentation was accomplished, thus dividing the news programs into different topics. They obtained reasonable results in their experiments, however their approach is completely dependent on the production rules of TV channels. Also aiming at achieving news story segmentation, Liu and Wang [16] focused their efforts on using a convolutional neural network in order to partition the programs into semantically meaningful parts. They based their input on the closed caption content of the news and trained and tested their model on TDT2 dataset, from Topic Detection and Tracking (TDT). Although they obtained significant results, their approach is limited to the linguistic information extracted from closed caption and thus not applicable to programs without such resource.

Even though several studies could be found, none comprises the same aspects and goals we aim at achieving with our study, which is to perform NLP and GIS extraction and structuring of stories depicted in TV news reports, focusing specially on urban issues complaints.

## III. The Crowd4City Geosocial Network

The Crowd4City system is a geosocial network aiming at providing e-participation to citizens, which enables them to take part more actively in their city's management, acting as sensors. The Crowd4City users can share and comment on many kinds of geolocated urban issues including traffic jam, criminality, potholes, broken pole lights and so on. Citizen's complaints on urban issues are shared publicly in the Crowd4City aiming to draw the attention from the authorities and the society as a whole. Hence, Crowd4City enables humans as sensors in a smart city environment. Figure 1 depicts the Crowd4City interface in which users can see the spatial distribution and pattern of different topics related to urban issues.

Regarding Crowd4City's use (Figure 1), the citizens can create complaint posts using their personal information or even anonymously, and they can input their dissatisfactions making use of the geographical tools. They can mark a single point

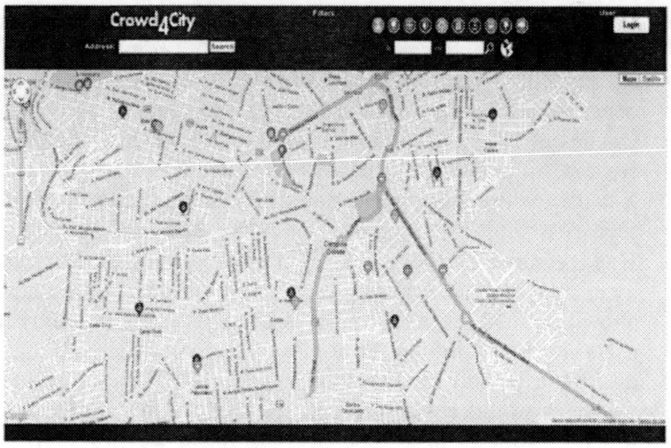


Figure 1. Crowd4City's main user interface.

on the map where the problem took place (for instance, if the user is reporting a pothole on a street); they can draw lines, perhaps to show routes where there are lighting issues; or they can even draw polygons on the map, thus being able to report regions that can be considered insecure.

Crowd4City presents some predefined categories for the problems reported including: Education, Sanitation, Transportation, Work Under Construction, Security and Others (Noise Pollution, Rubbish, Lighting, Potholes, etc.). However, if the user wishes to report something else, there is a category named "Other", which can be used for such uncategorized complaints.

Crowd4City's posts consist mainly of: location (geographic information), a title, a brief description and optionally multimedia attachments if the user has pictures or videos of the problem being reported. Additionally, the system provides a section for the other users's feedback with like/dislike buttons and a comment section, as seen on Figure 2.

Crowd4City enables operations such as pan and zoom. Also, the system made available several filters so that the users may perform more specialized searches for their information of interest. There are the basic filters, where the posts may be refined by the selected categories or creation dates; and the advanced filters, which may consider the complaints' contents and their geographic information. With the advanced filters,

Figure 2. A rubbish complaint.

the users may perform searches using the buffer and contains operations, may select some Points Of Interest (POI) categories (such as schools, hospitals, squares, airports and so on) and all the available filters may be used combined.

## IV. Automated Method for Extracting and Structuring Urban Issues Reported in TV News

The main problem addressed in this research deals with obtaining urban issues complaints from TV news, georeferencing them and automatically classifying them into one of the defined categories. The categories include sanitation, transportation, work under construction, among others. The urban issues context considered in this work is based on a corpus built in a previous work [8].

Our methodology comprises the following steps, according to Figure 3. First, we implemented a Web scraping method to extract the audio from video news. Second, we convert the audio into text using a speech recognition tool. Third, we use a gazetteer to perform geoparsing on the mentioned addresses and locations obtained from the Named Entity Recognition (NER) process, without preprocessing. Then, we implemented a preprocessing step comprising word capitalization, stopwords removal and lemmatization. Fourth, we use NER to obtain the named entities from the text. Then, we perform topic modeling to obtain the class of urban issues related to the text. Finally, the urban issues are located into the Crowd4City geosocial network. We detail each step of our methodology next.

### A. *Web scraping - Video 2 Txt*

Initially, we developed a Web scraping tool for obtaining the videos from TV news website. The data comes from a Brazilian TV broadcast website in Portuguese. We used the Selenium library [17] and YouTubeDL [18] to download the audio files from the video URLs that were stored in a JavaScript Object Notation (JSON) file. Then, we used the SpeechRecognition library [19] with the Google Speech Recognition API to convert audio into text. In order to decode the speech into text, groups of vectors are matched to one or more phonemes, which is a fundamental unit of speech. The SpeechRecognition library relies on modern speech recognition systems based on neural networks and Voice Activity Detectors (VADs). In addition, Google Speech Recognition API is free and supports Brazilian Portuguese language with good results.

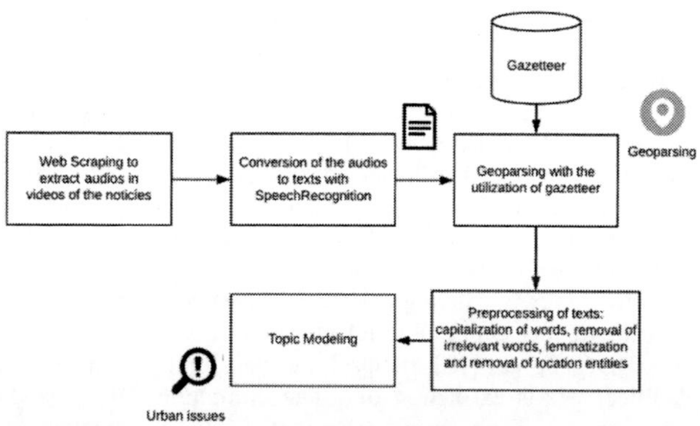


Figure 3. Our proposed methodology.

### *B. Preprocessing*

In the preprocessing step, we converted the text into lower case, removed stopwords and performed lemmatization. We used the Spacy library [20] to perform entity recognition of locations. The Natural Language Toolkit (NLTK) Python library [21] was also used for the lemmatization process.

The strategy defined for the preprocessing is to extract words that are entities from locations using Spacy NER, for which the library works very well when aided by the SpeechRecognition tool. Spacy also offers support for the Portuguese language, which avoids the translation of all texts into English, as it may reduce performance.

Spacy recognizes the location entities of the text and their title, so we combine all the location entities found to then search for those addresses and choose the one with the highest reliability.

We also have guaranteed anonymization by removing the names of people that took part in the audio extracted from video URLs. Hence, privacy was preserved, although it is important to note that all the videos processed in this work are publicly accessible from the sources.

### *C. Geoparsing*

We used the Geocoder library [22], that offers an API that enables the use of geocoding services such as Google, ArcGIS, and Bing. The chosen API service was ArcGIS, which provides simple and efficient tools for vector operations, geocoding, map creation and so on. Brazil is ranked level 1 in the library, which means that an address lookup will usually result in accurate matches to the "PointAddress" and "StreetAddress" levels, which fulfills our requirements. After having the entities properly combined, we iterate through this structure by checking which one is the most accurate address form the addresses the Geocoder returns using ArcGIS. The accuracy is increased by filtering the addresses found, so the user may perform filtering by state, city or even geographic coordinate.

We used Open Street Map (OSM) to obtain spatial data from some cities of the State of Paraiba in Brazil, and a gazetteer to improve the geoparsing accuracy. The gazetteer contains streets, neighborhoods, roads, schools, hospitals, supermarkets, pharmacy, etc. Notice that we do not deal with place names pronunciation, as the audio files do, because the names of the places are converted into text. We performed a cleaning of this data to keep only the information of interest to us: name, type and coordinates. Such cleaned data was stored in a PostgreSQL/PostGIS database system. Figure 4 presents the geoparsing step.

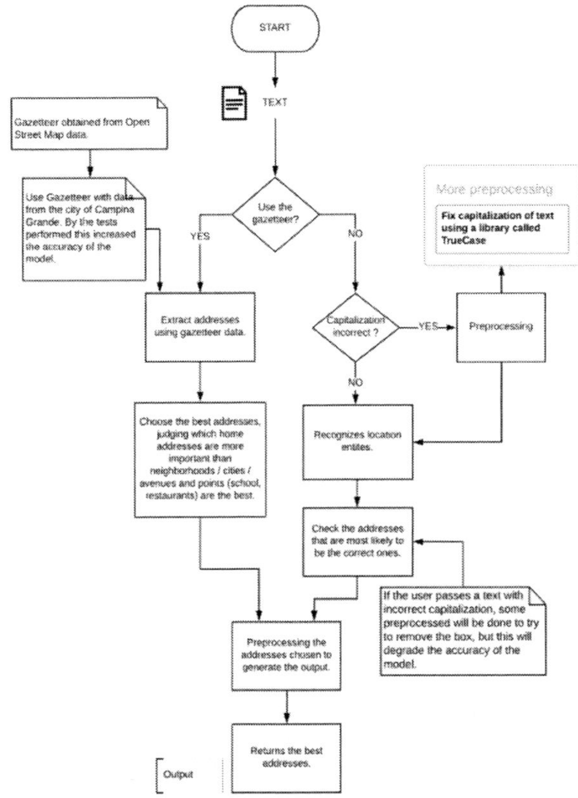


Figure 4. The geoparsing process.

### *D. Topic Modelling*

Concerning the topic modeling, we used Gensim [23], an open source library for unsupervised topic modeling and NLP, which provides statistical machine learning tools. We used LDA from Gensim (LDAMulticore and LDAModel) to implement topic modeling, which considers each document as a collection of topics, and each topic as a collection of keywords.

In order to implement a topic classifier in Gensim, we need to follow a few steps: creating both a word dictionary and a corpus (bag of words), then providing the desired number of topics and some algorithm tuning parameters. The word dictionary chooses an ID for all the words contained in documents, the corpus (bag of words) is a dictionary with word IDs and how many times that word repeats in the document. TF-IDF was also used, transforming the corpus co-occurrence matrix into a local TF-IDF co-occurrence matrix.

Concerning topic modeling, we removed all words that are location entities, as they are not useful for the class classification process, aiming at increasing the accuracy of the model. Thus, our classifier will focus on words of a given class, without worrying about locations.

To find out the best number of topics, some tests were performed and then it was verified which was the best model, comparing them with the measure coherence score, which evaluates the quality of the obtained topics. After these tests, we came to the conclusion that the best number of data topics would be four, as shown in Figure 5.

With four topics, the algorithm achieves a coherence score of 0.527613, the best result in the used dataset. Another improvement was to generate the 15 most repeated words in the topics generated by the algorithm. After that, we manually selected the words that should not be considered and we added them to the list of stop words. Then, we repeated the process until the 10 words in each topic were strongly related to the topic.

In topic modeling, we can analyze which topics represent all documents and also the keywords of each topic. Figure 6 shows the thirty most frequent words in the first topic and also presents the words of the first topic sorted according to their importance. The most important words are water, sewage,

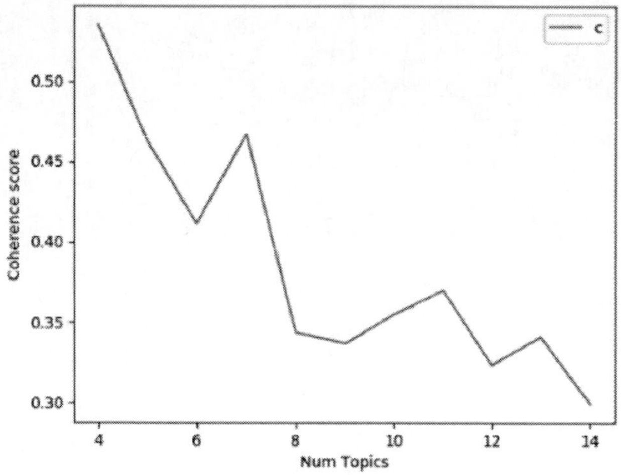


Figure 5. Coherence score per number of topics.

pavement, and home, thus indicating that the topic addresses sanitation problems.

## V. A CASE STUDY IN CAMPINA GRANDE NEIGHBOURHOOD

Usually, urban issues raised attention from local press, in order to establish a connection between population and city councils. In Campina Grande, a 400,000 inhabitants Brazilian city, there is a story in a local newspaper called "My Neighborhood on TV" that weekly shows existing urban issues and proposes to notify the authorities to solve a problem, defining a deadline to solve it. In general, citizens report their complaints to local TVs through messaging service platforms.

Thus, this research aims to fill the gaps mentioned above, helping to share complaints and providing a centralized means with this information, so it is easier for both the inhabitants to make complaints and for the local authorities to solve the reported problems.

### A. Setup

In this research, we collected 1,007 videos of the news story "My Neighborhood on TV", covering the years 2016 to 2019, with an average duration of five minutes per video. We took all videos from the Paraiba's TV news program website [24].

From all the videos obtained, in 602 of them (59.8 %) it was possible to get the text and locations. Unfortunately, some videos did not specify the location. At the same time, some videos did not specify the urban issue, as the word "obra", which means "something not concluded that is being built or repaired" in Portuguese, is applied as a problem generalization.

We extracted the problem classes reported in the videos, enabling various applications to use them in an attempt to improve city management, as the authorities can be notified and then provide solutions for urban issues in this easy manner.

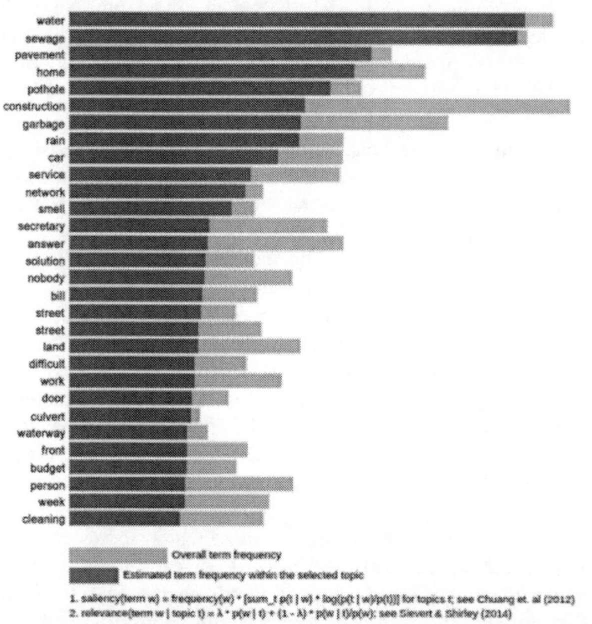


Figure 6. Most frequent words in the first topic.

### B. Results

We have performed several tests in the Gensim library, from changing pre-processing functions to changing parameters of the functions used. We used the number of steps equal to 10 because we saw that with this number we get good results without losing performance (see Figure 7). When trying to use values below or above 10, we saw that the accuracy began to decrease.

The metric we used to test the topics generated was the Coherence Score, which measures the relative distance between words within a topic. The number of parameters used was 4, with a score of 0.52. Such a score is acceptable in this preliminary study due to our dataset. We performed tests to verify how the generated model behaved with data not yet seen. One problem when using some geoparsing is in entity recognition. This is because the tools used for NLP cannot recognize entities that are misspelled (for example, if someone wrote the Campina Grande entity with all lowercase characters). However, this problem was mitigated with the use of the TrueCase library [25], which corrected the capitalization

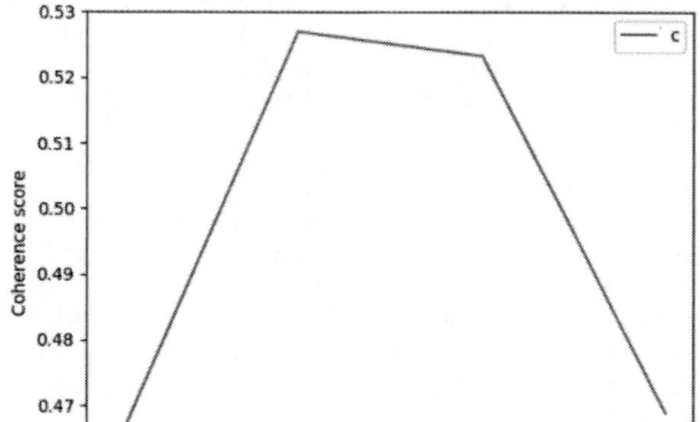


Figure 7. Comparative chart for influence of the steps value in the coherence score.

of words, so that the geoparsing used could recognize the entities, obtaining good accuracy. It is important to notice that NLP is by definition not capable of getting the real meaning of any term or context, as text is something by nature completely different than language. In order to deal with context, we need to combine NLP with other resources such as Part-Of-Speech tagging and supervised machine learning, for instance.

The TrueCase library supports the English language only. As in our case, the data was in Portuguese, hence we needed to use a library to translate the words from Portuguese to English - GoogleTrans [26] - use TrueCase and then do the reverse process, resulting in the words in Portuguese with the correct capitalization. However, sometimes, this procedure was unsuccessful due to problems in translations or problems in capitalized words.

As an additional process to improve the performance of geoparsing, we use a gazetteer, achieving improvements in the geolocation process for texts of the city of Campina Grande - which is the object of this research.

## VI. Conclusion

Citizens as sensors enable the engagement of society through technology to complain on urban issues. Smart cities demand tools for such engagement promoting e-citizenship and e-participation. Nonetheless, although some of such tools have already been proposed, it turns out that people engagement decrease in time. Hence, the obtention of urban issues from any media is very important to maintain people's engagement. As such, this paper proposes an approach to gather urban issues data from a TV news program and, using geoparsing and NLP techniques, to locate and classify the urban issues in order to input it in the Crowd4City geosocial network.

The results show that our approach is feasible and that we manage to classify urban issues into four topics: mobility, sanitation, buildings and others. As future work, we plan to perform an in-depth performance analysis of geoparsing, as well as topic modeling, by manually identifying the topics of the videos as ground truth and comparing them with the topic modeling results. Another plan consists of performing a comparative study between topic modeling and supervised machine learning.

## Acknowledgment

The authors would like to thank the Brazilian Research Council - CNPq for funding this research.

## References

[1] A. G. R. Falcão et al., "Towards a reputation model applied to geosocial networks: a case study on crowd4city," in Proceedings of the 33rd Annual ACM Symposium on Applied Computing, SAC, Pau, France, 2018, pp. 1756–1763.

[2] T. Wandhofer, C. van Eeckhaute, S. Taylor, and M. Fernandez, "WeGov analysis tools to connect policy makers with citizens online," in Proceedings of the tGovernment Workshop, 2012, pp. 1–7.

[3] N. Walravens, "Validating a Business Model Framework for Smart City Services: The Case of FixMyStreet," in Proceedings of the 27th International Conference on Advanced Information Networking and Applications Workshops, 2013, pp. 1355–1360.

[4] D. Britz, A. Goldie, M.-T. Luong, and Q. V. Le, "Massive exploration of neural machine translation architectures," ArXiv, vol. abs/1703.03906, 2017.

[5] S. Reddy, D. Chen, and C. D. Manning, "Coqa: A conversational question answering challenge," vol. Transactions of the Association for Computational Linguistics, Volume 7, March 2019, pp. 249–266. [Online]. Available: https://www.aclweb.org/anthology/Q19-1016 [accessed: 2020-03-02]

[6] H. Schwenk, L. Barrault, A. Conneau, and Y. LeCun, "Very deep convolutional networks for text classification," in EACL (1), M. Lapata, P. Blunsom, and A. Koller, Eds. Association for Computational Linguistics, 2017, pp. 1107–1116. [Online]. Available: https://www.aclweb.org/anthology/E17-1104/ [accessed: 2020-03-02]

[7] A. Radford et al., "Language models are unsupervised multitask learners," 2018. [Online]. Available: https://d4mucfpksywv.cloudfront.net/better-language-models/language-models.pdf [accessed: 2020-03-02]

[8] M. G. de Oliveira, C. de Souza Baptista, C. E. C. Campelo, and M. Bertolotto, "A Gold-standard Social Media Corpus for Urban Issues," in Proceedings of the Symposium on Applied Computing (SAC), ser. SAC '17. New York, NY, USA: ACM, 2017, pp. 1011–1016.

[9] N. Camelin et al., "FrNewsLink : a corpus linking TV Broadcast News Segments and Press Articles," in Proceedings of the Eleventh International Conference on Language Resources and Evaluation (LREC 2018). Miyazaki, Japan: European Language Resources Association (ELRA), 2018. [Online]. Available: https://www.aclweb.org/anthology/L18-1329 [accessed: 2020-03-02]

[10] R. Kannao and P. Guha, "Overlay Text Extraction From TV News Broadcast," CoRR, vol. abs/1604.00470, 2016. [Online]. Available: http://arxiv.org/abs/1604.00470 [accessed: 2020-03-02]

[11] M. Pala, L. Parayitam, and V. Appala, "Real-time transcription, keyword spotting, archival and retrieval for telugu TV news using ASR," International Journal of Speech Technology, vol. 22, no. 2, 2019, pp. 433–439.

[12] R. Bansal and S. Chakraborty, "Visual Content Based Video Retrieval on Natural Language Queries," in Proceedings of the 34th ACM/SIGAPP Symposium on Applied Computing, ser. SAC '19. New York, NY, USA: ACM, 2019, pp. 212–219.

[13] Z. Dong and X. Lv, "Subject extraction method of urban complaint data," in Proceedings of the IEEE International Conference on Big Knowledge (ICBK), 2017, pp. 179–182.

[14] B. Mocanu, R. Tapu, and T. Zaharia, "Automatic extraction of story units from TV news," in Proceedings of the IEEE International Conference on Consumer Electronics (ICCE), Jan 2017, pp. 414–415.

[15] T. Zlitni, B. Bouaziz, and W. Mahdi, "Automatic topics segmentation for TV news video using prior knowledge," Multimedia Tools and Applications, vol. 75, no. 10, 2016, pp. 5645–5672.

[16] Z. Liu and Y. Wang, "TV News Story Segmentation Using Deep Neural Network," in Proceedings of the IEEE International Conference on Multimedia Expo Workshops (ICMEW), 2018, pp. 1–4.

[17] Selenium, "Selenium Library." [Online]. Available: https://www.selenium.dev/ [accessed: 2020-03-02]

[18] R. Gonzalez et al., "YouTubeDL." [Online]. Available: https://github.com/ytdl-org/youtube-dl [accessed: 2020-03-02]

[19] A. Zhang, "Selenium." [Online]. Available: https://github.com/Uberi/speechrecognition [accessed: 2020-03-02]

[20] Explosion AI, "Spacy." [Online]. Available: https://spacy.io/ [accessed: 2020-03-02]

[21] NLTK Project, "The Natural Language Toolkit." [Online]. Available: https://radimrehurek.com/gensim/ [accessed: 2020-03-02]

[22] D. Carriere et al., "Geocoder." [Online]. Available: https://geocoder.readthedocs.io/ [accessed: 2020-03-02]

[23] R. Řehůřek and P. Sojka, "Software Framework for Topic Modelling with Large Corpora." [Online]. Available: https://radimrehurek.com/gensim/ [accessed: 2020-03-02]

[24] G1 Paraiba, "JPB1 TV News program official website." [Online]. Available: http://g1.globo.com/pb/paraiba/jpb-1edicao/videos/ [accessed: 2020-03-02]

[25] D. Fury, "TrueCase." [Online]. Available: https://github.com/daltonfury42/truecase [accessed: 2020-03-02]

[26] S. Han, "Googletrans." [Online]. Available: https://github.com/ssut/py-googletrans [accessed: 2020-03-02]

# Using Satellite Imagery and Vegetation Indices to Monitor and Quantify the Performance of Different Varieties of Camelina Sativa

Mar Parra[(1)], Lorena Parra[(1,2)], David Mostaza-Colado[(2)], Pedro Mauri[(2)], Jaime Lloret[(1)]

[(1)] Instituto de Investigación para la Gestión Integrada de Zonas Costeras, Universitat Politècnica de València C/ Paranimf nº 1, Grao de Gandía – Gandía, Valencia, Spain

[(2)] Instituto Madrileño de Investigación y Desarrollo Rural, Agrario y Alimentario (IMIDRA), Finca "El Encin", A-2, Km 38, 2, 28800 Alcalá de Henares, Madrid, Spain

E-mail: maparbo@epsg.upv.es, loparbo@doctor.upv.es, david.mostaza@madrid.org, pedro.mauri@madrid.org, jlloret@dcom.upv.es

***Abstract*—In recent years, the cropping of *Camelina sativa* has gained popularity among the farmers of rainfed crops. It is an annual and flexible crop, which can grow in different regions. The estimate of the crop yield is essential for farmers. *Camelina sativa* is a small plant that forms a uniform green tapestry of grass. Hence, satellite imagery can be used for monitoring the crops. In this study, we present the use of Sentinel-2 data to monitor the performance of 6 varieties of *Camelina sativa*. Crops have been growing from fall to spring, and the harvest occurred in early June. We include satellite imagery from February to June. In this paper, we include a single image per month. Moreover, due to the size of the plots, we only consider the data from bands with a spatial resolution of 10m and 20m. First of all, the differences in spectral signatures of varieties along the time are presented. Then, we detail the possibilities of correlation between different vegetation indices and crop harvest. Finally, a multivariable statistical analysis to correlate bands of Sentinel-2 with harvested seeds is shown. This analysis estimates the yield with high accuracy.**

***Keywords-Sentinel-2; rainfed crops; multivariable statistical analysis; NDWI; NDMI; EVI.***

## I. Introduction

Intensive agriculture is vital in our modern societies to produce enough food to sustain the ever-growing population. These cultures are too big to be managed in the same way traditional cultures have. Nevertheless, they do need to be monitored to obtain peak productivity and performance. The consequences of a large estate not being adequately handled are proportional to the magnitude of the field. The bigger the area is, the more losses it will experience with low performance. Therefore, an urgent need for the development of a monitoring system for intensive agriculture has arisen.

Nowadays, the most used method for agricultural monitoring is the use of Wireless Sensor Networks (WSN). Their purpose is focused on monitoring the soil and the chemical characteristics of the plants, such as nitrogen content, though. Instead, our proposed method measures the yield. Unmanned Aerial Vehicles (UAVs) have been proved to be helpful devices for Geographic Information System (GIS) [1]. Environmental variables, such as tree coverage, can be monitored with the use of imaging techniques. The process of obtaining these images was done first by hand until the introduction of UAVs. This innovation allows for these surveys to be done remotely.

The use of airborne multispectral and hyperspectral imagery and high-resolution satellite imagery has been proved to be useful. Moreover, other imaging analysis techniques have been tested lately [2]. A study developed this year managed to detect fruits in trees using image processing [3].

The application of new monitoring techniques to manage intensive agriculture is critical. Not only would it mean diminishing the use of our resources, but it would also translate in an improvement of the yield. Moreover, monitoring the productivity of the crop would help detect problems. It is possible that we should be having more yield than the harvested one. That would mean something is preventing it from being as productive as it should. Besides, estimating the yield has some economic benefits. The cost-benefit ratio could be calculated before the crops are harvested. This could also mean knowing the price at which the product could be sold before other companies and being able to prepare in advance.

The aim of this paper is to determine if the multispectral imaging data, which is obtained from the satellite Sentinel-2B and Sentinel-2A, can be used to determine a key parameter in agricultural productivity and performance. The said parameter is the number of seeds several *Camelina sativa* crops produce, using up to six different varieties. The *Camelina sativa* is a crop that is currently being sown in many dry areas of the world. The plants produce seeds that are used for oil extraction. This plant from the Brassicaceae family is annual, which makes its monitoring easier. It creates a fruit that contains up to sixteen seeds, according to Mostaza-Colado et al. [4]. In order to accomplish our goal, we will obtain the spectral signature of the crops using images taken once per month from February to July. The images used will be taken at the end of each month with the last one representing the bare ground after the seeds are collected. Several vegetation indices will be analyzed using the information from these images to try and relate them with productivity. If necessary, we will create our indices using statistics.

The rest of the paper is structured as follows. The discussion of the related work is presented in Section 2. Section 3 deals with the materials and methods that were used for this experiment. The results are portrayed in Section 4. Finally, Section 5 shows the conclusions of this work.

## II. Related Work

In this section, we discuss some papers which deal with different methods to monitor crops. Moreover, other different vegetation indices used nowdays, which could be useful for our purpose, are mentioned.

Mostaza-Colado et al. [4] performed preliminary tests to check if Sentinel-2B images could be used to estimate the growth of *Camelina sativa*. They attempted to correlate the Normalized Difference Vegetation Index (NDVI) with the growth of the plant. They proved a correlation between the acquisition techniques. However, they could not prove the existence of a relationship between the NDVI and the yield.

Sankey et al. [5] used a UAV equipped with a Light Detection and Ranging (LiDAR) sensor, as well as hyperspectral imaging, to monitor a forest in the southwest of the USA. They determined that the data could be analyzed to generate 3D point cloud data, although the differences between ground and trees were not evident in the dense parts of the forest. This is not a problem in the case of intensive cultures, where the coverage is never as thick as a forest.

Vega et al. [6] used a UAV to monitor a sunflower crop and determined that their method could be used in precision agriculture. They managed to extract the NDVI from the images. Moreover, they correlated the NDVI with aerial biomass, plant nitrogen, and grain yield. One of the advantages they remarked from UAVs compared to satellites is the ability to obtain images on cloudy days.

Ashtekar et al. [7] attempted to map the surface water dynamics in the upper Krishna River basin. To do so, they modeled the water dynamics using the Normalized Difference Water Index (NDWI), taking data from 17 years. This index allowed them to classify the water as permanent, seasonal, and new permanent.

A study similar to the one we propose was developed by Yawata et al. [8]. They used satellite images to extract the spectral values and then estimated the rice yield employing a mixed model. Two vegetation indices were implemented as feature values: the NDVI and the Green Normalized Difference Vegetation Index (GNDVI). They managed to reduce the mean absolute error compared to other estimation methods, such as regression methods.

Selbmann et al. [9] used several indices derived from Landsat imaging to monitor wildfire consequences in a wetland tundra ecosystem. The indices they used were the NDVI, the Enhanced Vegetation Index (EVI), the Normalized Difference Moisture Index (NDMI), and the Normalized Burn Ratio (NBR). They managed to relate the EVI and NDVI with the severity of the fires.

Fassnacht et al. [10] attempted to develop a non-destructive method to estimate the carotenoid content on trees. They used the Angular Vegetation Index (AVI) to do so. Said index had to be combined with two other proposed carotenoid indices to give an accurate enough output.

The performance of corn crop fields was estimated by Venancio et al. [11] using the FAO-66 approach and the Soil Adjusted Vegetation Index (SAVI). They used the seventh and eight bands from Landsat to forecast the corn yield at the farm-level in Brazil. The predictions they obtained showed little difference from the real value (between -5% and 5%).

Marin et al. [12] managed to determine the grass coverage in urban lawns with RGB histograms of the lawns. Brightness values between 40 and 60 extracted from the green layer could be used to determine the coverage.

Among the studies mentioned above, several have a similar objective to the one we have. Nevertheless, they used already existing indices while we will test new combinations. Moreover, in our experiment, we will be using *Camelina sativa*, which is an emerging crop. Furthermore, we will be using images from several months. In conclusion, we will estimate the productivity of a *Camelina sativa* crop using geoprocessing, no matter its variety. This will be done using satellite imaging. In order to obtain the desired results, we will compare the value of the bands using several vegetation indices. Moreover, we will create our indices if necessary.

## III. MATERIAL AND METHODS

In this section, the utilized images, how they were obtained, and the methodology applied are described.

### *A. Image obtention*

Among the available open-access images from the different satellites, we have selected to work with data from Sentinel-2. The Sentinel-2 was chosen due to its high spectral resolution, up to 12 bands (B), and four generated indices. Furthermore, it has a high spatial resolution, which is 10m (for B2, B3, B4, and B8), 20m (B5, B6, B7, B8A, B11, and B12), and 60m (B1, B9, and B10). Besides, this satellite presents a high temporal resolution, which allows having one set of data every five days. Other satellites that offer open-access images give a lower temporal, spatial, and spectral resolution.

The images are obtained from the Copernicus Open Access Hub webpage [13]. The studied plots *with Camelina sativa* are located in the T30TVK of the grid system. All the images obtained between January and June of 2019 are downloaded. However, the data from January is not used due to the vegetation not yet being visible. First, we discard the images with cloud coverage in the studied area. Next, we select the date of the pictures to have the first picture at the end of February, have all the images separated by 30 days (average), and without cloud coverage. Therefore, the images used correspond to 28-February, 30-March, 29-April, 29-May, and 30-June. The first four images will show the changes in the vegetation, while the last picture will represent the soil status after the harvest.

### *B. Data gathering*

Once the satellite imagery has been obtained and selected, the next step is to get the values of the pixel of different bands for the different *Camelina sativa* varieties. The plots were already digitalized in previous studies [4].

Thus, using ArcMap [14], the satellite imagery and the digitalized plots are opened, see Figure 1. In this figure, we show the plots, identified in yellow borders, and the area which is considered for statistical analysis indicated in red.

This area is smaller to avoid the effect of adjacent surfaces. Every plot contains a single variety of *Camelina sativa*.

The tool Zonal Stadistic as a Table [15] is selected to obtain the values of each band for the different plots. All the statistics are obtained for every band. These data are then exported to an Excel file. In Excel, we generate the different spectral signature for each variety along the time using the mean and median values, including the standard deviation calculated in the previous step. The included varieties in this study were obtained from the Camelina Company España [16]. Purchased seeds were sowed at the beginning of December. The varieties are named 1), 2) 3), 4), 5), and 11).

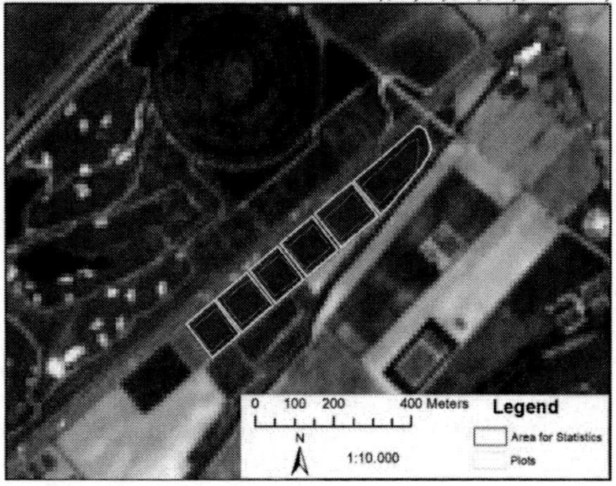


Figure 1. Studied area.

## C. Vegetation indices calculation

To obtain a correlation with the harvested seeds, we use the data from the band 1 (B1) to band 12 (B12) of each image to calculate different vegetation indices. Now, we are going to define the utilized indices. Previous works [4] have evaluated the NDVI of these plots. They did not find any relation between NDVI and harvested quantities. Thus, in this paper, we increase the evaluated indices and include the following:

(i) NDWI [17], which is based on the green and Short Wave Infrared (SWIR) bands. In the case of Sentinel-2, the formula to calculate the index is (B3-B8)/(B3+B8). NDWI gives information about the water content in the plants. This index can have values between -1 and 1. The lower the value, the greater the water content. The higher the value, the lower the vegetation cover and water content.

(ii) NDMI [18], which is based on the Near Infrared (NIR) and SWIR. In Sentinel, the formula is (B8–B11)/(B8+B11). This index offers information about vegetation water content. As the previous index, this one can adopt values between -1 and 1. The higher the value, the lower the water stress.

(iii) EVI [19], which was developed by NASA as an alternative to NDVI and similar indices. This index has two main advantages over NDVI-like indices: (i) more sensitive in areas with high biomass and (ii) reduces the influence of atmospheric conditions. It is calculated using the B2, B4, and B8. Besides, some constants are used. The formula is 2.5*((B8A-B4)/((B8A+6*B4-7.5*B2)+1)). In contrast to previous indices, this one is not limited to values from -1 to 1.

## D. Obtaining new correlations

Finally, we perform multivariate analysis to find a possible association between different bands and indices and the harvested seeds of the different varieties. The main objective is to have a preliminary result that indicates any potential band or bands for its future use when creating a vegetation index that predicts the harvest. In the case that any band presents a correlation with the harvest, we will use regression tools to define this correlation.

# IV. RESULTS

In this section, we discuss the results of this contribution. First, the differences in the spectral signatures are detailed. Next, we present the analysis of the indices. Finally, the multivariate analysis and its outcomes are discussed.

## A. Spectral signatures

After obtaining the satellite imagery, some problems were detected. First of all, in images from January to April, the data of band ten was missing. Furthermore, in images from May and June, the data from the calculated indices corresponding to the Level 2A specific bands: Scene-average Water Vapour map (WVP), Aerosol Optical Thickness map (AOT), and Scene Classification (SCL) were not included. Therefore, for the analysis of spectral signatures, the data is not complete for all the time-series. Moreover, we only include the data with a spatial resolution of 10 and 20m. This data can be seen in Figure 2; the name in brackets indicates the variety of *Camelina sativa*. In this figure, the mean value of pixel for each band in different moments of the year is displayed. The months, February to June, are represented in different colors and indicated as 2 to 6. The colors are to show the phenological conditions of the crop. In green, we describe the moments when the *Camelina sativa* has a green coloration and is growing. In yellow, we indicate the period in which plants have very low water content, and they are dry. In late June, the assigned color is brown because the plants are completely dry, and the seeds are already collected.

The first thing that can be noticed when analyzing Figure 2 is that different varieties seem to have different patterns. This might be caused by differences in the phenological characteristics of the different species. Next, we present in detail some of these differences. For example, varieties 3) and 4) present higher variations in the red band between February and March than 1) and 2) (which do not show any change) or 5) and 11) (which decrease to a lesser extent). From March to April, most of the varieties increase their mean value in the green band, nonetheless 5) experiences a decrease.

Apart from that, there are some changes in the region of 705 to 783nm (IR light), which is commonly used for vegetation characterization. While for most of the varieties, during the moment in which plants are green, the minimum values in those bands are found in March, 2) has similar data in bands 6 to 8 in February and March. Taking into account that all the plants were sowed at the same moment, the soil was homogenized, and the environmental conditions were the same, the differences found are due to the different varieties. Thus, spectral signatures can be used to characterize the crops.

### *B. Vegetation Indices*

The indices above are calculated for the different varieties and different periods of the year.

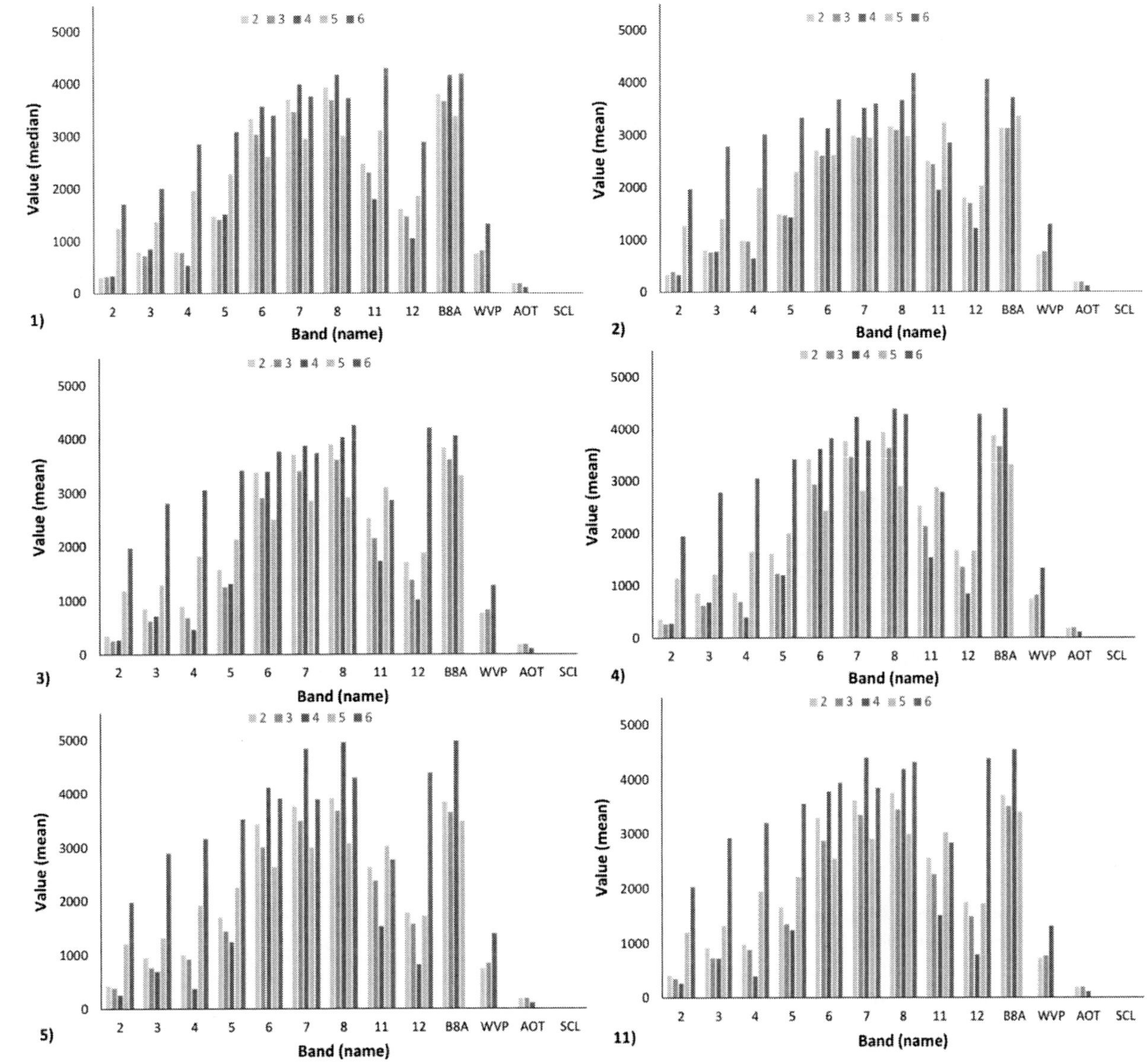


Figure 2. Spectral signatures for different varieties and periods (February to June).

Data from index NDWI is displayed in Table I. This index indicates the changes in the water content of the surface, in this case, the crop. The results of the NDMI index are presented in Table II. The NDMI is an indication of the moisture. For Tables I and II, the varieties were ordered according to the harvested amount of seeds. The variety 5) (620kg/Ha) is the one with the lowest harvest and variety 3) is the one with the highest harvest (1125kg/Ha). The other varieties have a harvest between 914 and 971 kg/Ha. Both indices are similar, and the results of their application offer identical data. The results point out that the healthiest crops are 3), 4), 5), and 11). Nevertheless, there is no relation between the results of the index and the harvested quantity.

Next, the results of EVI are presented in Table III. The higher the value of the index is, the higher the plant vigor is.

According to EVI, the healthiest crops are 5), 4), and 11). Again, no relation was found between the index and the harvested seeds of different varieties.

TABLE I. VALUES OF NDWI FOR DIFFERENT VARIETIES

| Month | NDWI per Varieties | | | | | |
|---|---|---|---|---|---|---|
| | *5)* | *2)* | *1)* | *4)* | *11)* | *3)* |
| 2 | -0.61 | -0.66 | -0.66 | -0.65 | -0.61 | -0.64 |
| 3 | -0.66 | -0.65 | -0.67 | -0.71 | -0.65 | -0.71 |
| 4 | -0.76 | -0.68 | -0.66 | -0.73 | -0.72 | -0.70 |
| 5 | -0.40 | -0.36 | -0.37 | -0.41 | -0.39 | -0.39 |
| 6 | -0.33 | -0.31 | -0.30 | -0.32 | -0.31 | -0.31 |

TABLE II. VALUES OF NDMI FOR DIFFERENT VARIETIES

| Month | NDMI per Varieties | | | | | |
|---|---|---|---|---|---|---|
| | *5)* | *2)* | *1)* | *4)* | *11)* | *3)* |
| 2 | 0.20 | 0.12 | 0.23 | 0.22 | 0.19 | 0.21 |
| 3 | 0.21 | 0.12 | 0.23 | 0.26 | 0.21 | 0.25 |
| 4 | 0.53 | 0.31 | 0.40 | 0.48 | 0.49 | 0.40 |
| 5 | 0.01 | -0.04 | -0.02 | 0.00 | 0.00 | -0.03 |
| 6 | -0.05 | -0.07 | -0.07 | -0.06 | -0.06 | -0.07 |

TABLE III. VALUES OF EVI FOR DIFFERENT VARIETIES

| Month | EVI per Varieties | | | | | |
|---|---|---|---|---|---|---|
| | *5)* | *2)* | *1)* | *4)* | *11)* | *3)* |
| 2 | 1.08 | 0.84 | 1.23 | 1.18 | 1.06 | 1.14 |
| 3 | 1.09 | 0.88 | 1.24 | 1.27 | 1.06 | 1.28 |
| 4 | 2.18 | 1.49 | 1.86 | 2.13 | 2.12 | 1.87 |
| 5 | 0.51 | 0.46 | 0.48 | 0.75 | 0.47 | 0.56 |
| 6 | 0.29 | 0.27 | 0.27 | 0.31 | 0.27 | 0.30 |

Consequently, we can affirm that there is no correlation between different tested indices and harvested seeds. Thus, the indices cannot be used for the prediction of harvest.

### *C. Correlation of bands and the harvest*

Since none of the typical vegetation indices tested in this paper and a previous one [4] has offered a correlation with the harvest, we will perform a multivariate analysis to find a relationship. In this analysis, we will include the harvest quantity in kg/Ha of the six varieties and the value of the included bands in this paper (from February to June).

A multivariate analysis with up to 95 variables (16 bands + 3 indices per 5 months, and the harvest) is conducted with Statgraphics Centurion. The results of the analysis indicate that two bands are correlated with the harvest. The first band is the WVP of April and the second one is on the B1 of June. The one that presents a higher correlation and is meaningful in terms of prediction is the WVP of April (WVP4). According to Statgraphics, the p-value of that correlation is 0.0117, and the correlation coefficient is -0.9103. Figure 3 shows the correlation between the three WVP analyzed and the harvest. There, we can see the correlation that exists among harvest and WVP.

The last step is to perform a simple regression with Statgraphics to obtain a mathematical model that correlates both variables. First, we verify the comparison of regression models available in the software. The one that offers a higher correlation is the reciprocal-Y squared-X model. The graphic that shows this correlation, the mathematical model, and the intervals (prediction and confidence) are shown in Figure 4.

The mathematical equation of this model is Eq (1); its correlation coefficient is 0.926, and the squared-R 85.81. The standard error is 0.0001 and the mean absolute error is 0.00007. Finally, the p-value of the model is 0.0079. All this data confirms that the model is accurate and it can be used to predict in the future the harvest of *Camelina sativa* crops based on data of WVP.

$$\textbf{Harvest} = 1/(-0.00229478 + 1.95954E\text{-}9 * WVP4^{2}) \quad (1)$$

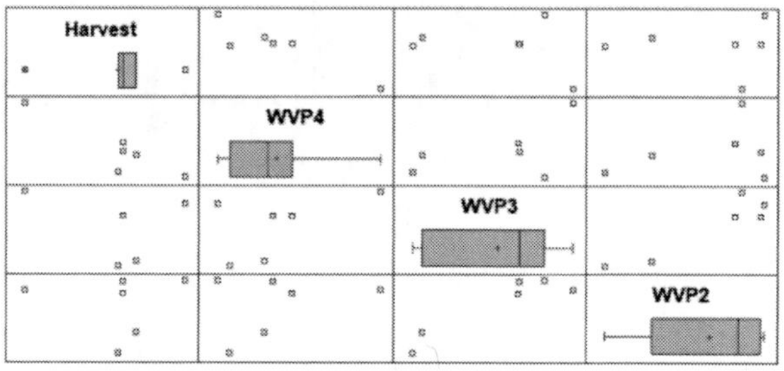


Figure 3. Spectral signatures for different varieties and periods (February (2) to April (4)).

It should be considered that the different varieties of *Camelina sativa* present different seeds size. According to [16], varieties such as 3) and 5) are the ones that give the bigger seeds, and 36 is the one that presents the higher height and larger inflorescences. It is possible that due to the characteristics of the seeds, some have been lost before the harvest because of the wind and other adverse meteorological conditions. On the other hand, 1), 3) and 6) have a smaller size and the loss of seeds due to the wind is minimized. It must be considered that the decision of the harvesting moment is crucial in order to reduce seed loss. This loss might have an impact on the prediction model presented in this paper.

## V. CONCLUSION

In this paper, we present the use of satellite imagery for the monitoring of different varieties of *Camelina sativa*. According to the results of the spectral signatures, we identify a different phenology in different varieties. They have different patterns in visible and IR bands. We calculate NDWI, NDMI, and EVI indices to find a possible correlation between indices and harvest. None of the typical vegetation indices tested in this paper present a correlation.

Nevertheless, a multivariate analysis was carried out with Statgraphics. The results point out that the WVP4 is correlated with the harvest. The regression model was obtained with a correlation coefficient of 0. 926. Thus, we have demonstrated the usefulness of the satellite imagery for *Camelina sativa* monitoring and harvest prediction.

For future work, we will extend our study and include more images to evaluate the best moment for WVP measurement to have a more accurate model. Moreover, to avoid the disturbances of clouds and other atmospheric factors, the utility of images obtained with a drone with a thermal camera will be evaluated. Furthermore, the study would be run for several years to eliminate a possible year-specific effect.

## ACKNOWLEDGMENT

This work has been partially supported by European Union through the ERANETMED (Euromediterranean Cooperation through ERANET joint activities and beyond) project ERANETMED3-227 SMARTWATIR and by the Conselleria de Educación, Cultura y Deporte with the Subvenciones para la contratación de personal investigador en fase postdoctoral, grant number APOSTD/2019/04, and by "Fondo Europeo Agrícola de Desarrollo Rural (FEADER) – Europa invierte en zonas rurales", the MAPAMA, and Comunidad de Madrid with the IMIDRA, under the mark of the PDR-CM 2014-2020" project number PDR18- CAMEVAR.

## REFERENCES

[1] B. Bollard-Breen et al., "Application of an unmanned aerial vehicle in spatial mapping of terrestrial biology and human disturbance in the McMurdo Dry Valleys, East Antarctica," *Polar Biology*, vol. 38, no. 4, pp. 573-578, April 2015.

[2] B. Basnet and J. Bang, "The state-of-the-art of knowledge-intensive agriculture: a review on applied sensing systems and data analytics," *Journal of Sensors*, vol. 2018, article ID 3528296, September 2018.

[3] L. García et al., "Quantifying the Production of Fruit-Bearing Trees Using Image Processing Techniques," INNOV 2019, The Eighth International Conference on Communications, Computation, Networks and Technologies, 24-28 November, Valencia, Spain, 2019

[4] D. Mostaza-Colado, P. V. Mauri Ablanque, and A. Capuano, "Assessing the Yield of a Multi-varieties Crop of Camelina sativa (L.) Crantz through NDVI Remote Sensing," 2019 Sixth International Conference on Internet of Things: Systems, Management and Security (IOTSMS), 22-25 October, Granada, Spain, 2019. pp. 596-602

[5] F. Neugirg et al., "Erosion processes in calanchi in the Upper Orcia Valley, Southern Tuscany, Italy based on multitemporal high-resolution terrestrial LiDAR and UAV surveys," *Geomorphology*, vol. 269, pp. 8—2, September 2016.

[6] F. Agüera Vega, F. Carvajal Ramírez, M. Pérez Saiz, and F. Orgaz Rosúa, "Multi-temporal imaging using an unmanned aerial vehicle for monitoring a sunflower crop," *Biosystems Engineering*, vol. 132, pp. 19-27, April 2015.

[7] A. S. Ashtekar, M. A. Mohammed-Aslam, and A. R. Moosvi, "Utility of Normalized Difference Water Index and GIS for Mapping Surface Water Dynamics in Sub-Upper Krishna Basin," *Journal of the Indian Society of Remote Sensing*, vol. 47, no. 8, pp. 1431-1442, August 2019.

[8] K. Yawata, T. Yamamoto, N. Hashimoto, R. Ishida, and H. Yoshikawa, "Mixed model estimation of rice yield based on NDVI and GNDVI using a satellite image," Remote Sensing for Agriculture, Ecosystems, and Hydrology XXI, 9-12 September, Strasbourg, France, 2019.

[9] A. K. Selbmann, M. M. Loranty, S. Natali., and M. Wegmann, "Assessment of wildfire severity and vegetation recovery in tundra ecosystems using time series of satellite-derived vegetation indices from the Yukon-Kuskokwim-Delta, Alaska," American Geophysical Union, Fall Meeting 2018, December 2018.

[10] F. E. Fassnacht, S. Stenzel, and A. A. Gitelson, "Non-destructive estimation of foliar carotenoid content of tree species using merged vegetation indices," *Journal of Plant Physiology*, vol. 176, pp. 210-217, March 2015.

[11] L. Venancio et al., "Forecasting corn yield at the farm level in Brazil based on the FAO-66 approach and soil-adjusted vegetation index (SAVI)," *Agricultural Water Management*, vol. 225, November 2019.

[12] J. Marín, J. Rocher, L. Parra, S. Sendra, J. Lloret, and P. V. Mauri, "Autonomous WSN for Lawns Monitoring in Smart Cities," 2017 IEEE/ACS 14th International Conference on Computer Systems and Applications (AICCSA), 30 October-03 November, Hammamet, Tunisia, 2017.

[13] Copernicus Open Access Hub Webpage. Available at: https://scihub.copernicus.eu. Last Access on 04/12/2019

[14] ArcGIS Desktop ArcMap. Available at: https://desktop.arcgis.com/en/arcmap/. Last Access on 09/03/2020

[15] ArcGIS Desktop 9.3 Help: Zonal Statistics as Table. Available at: http://webhelp.esri.com/arcgisdesktop/9.3/index.cfm?topicname=Zonal_Statistics_as_Table. Last Access on 09/03/2020

[16] Camelina Company España Webpage – Varieties of Camelina. Available at: http://camelinacompany.es/variedades/. Last Access on 04/12/2019

[17] S. K. Mcfeeters, "The use of the Normalized Difference Water Index (NDWI) in the delineation of open water features", International journal of remote sensing, vol. 17, no. 7, pp. 1425-1432, 1996.

[18] B. C. Gao, " NDWI—A normalized difference water index for remote sensing of vegetation liquid water from space", Remote sensing of environment, vol. 58, no. 3, pp. 257-266, 1996.

[19] W. J. Van Leeuwen, A. R. Huete, and T. W. Laing, "MODIS vegetation index compositing approach: A prototype with AVHRR data", Remote Sensing of Environment, vol. 69, no. 3, pp. 264-280, 1999.

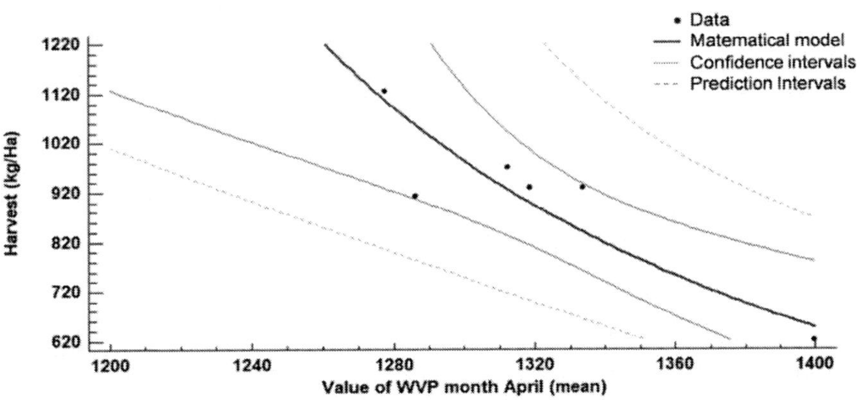


Figure 4. Spectral signatures for different varieties and periods (February to June).

# A Tool for Spatially Based Prediction of Consumer Lawsuits against Electric Power Companies

Domingos A. Dias Junior, Johnatan C. Souza, João O. B. Diniz, Geraldo Braz Junior, João D. S. Almeida, Anselmo Cardoso de Paiva
Applied Computing Group (NCA)
Federal University of Maranhão (UFMA)
São Luís Brazil
Email: {domingos.adj; johnatancarvalho; joao.bandeira; geraldo; jdallyson; paiva}@nca.ufma.br

Erika W. B. A. L. Alves
Equatorial Energy, Brazil
São Luís, Brazil
Email: erika.assis@equatorialenergia.com.br

*Abstract*—**The main purpose of an energy company is the provision of services to the final consumer. This does not mean that completely avoiding failures of the system is an easy task. These failures could lead to problems in the relationship between the energy company and the clients, resulting in judicial dispute. Thus, it is interesting for a power company to preemptively identify the consumers who are dissatisfied. Therefore, it is important for company executives to identify regions or groups of consumers that are related to the same cause that motivated a lawsuit. Thus, the present work aims to propose a tool for spatial analysis and spatially based prediction of consumer lawsuits against an electric power company using data mining and machine learning techniques. The results obtained from a database of an electric power company in Brazil showed results with 96.52% sensitivity in identifying consumers with lawsuits related to Unregistered Power Consumption (UPC). Also, a tool for visualization and spatial analysis of this group of clients is presented.**

***Keywords - lawsuit; electricity sector; consumer profile; geoanalysis; geovisualization.***

## I. INTRODUCTION

Nowadays, many researchers have been studying problems in large volumes of data to analyze the consumer profile, using the concepts of both data mining and machine learning. We can define data mining as the process of exploring large amounts of data looking for consistent patterns. One way to find consistent patterns and associations within databases is to use machine learning techniques [2].

The energy market in several countries is changing, basically due to the deregulation of the market and the emergence of judicial and administrative mechanisms for consumer protection [3]. The main purpose of an energy company is to provide services to the final consumer, even though it is very difficult to completely eliminate failures of the system [4]. System failures could result in a judicial dispute between the customer and the energy company. For that reason, it is useful for a power company to preemptively identify the consumers who are dissatisfied and also to identify the reasons of this in order to generate action plans to avoid lawsuits and increase the quality of service. Thus, there is a need for a prediction service for lawsuits.

By analyzing large data in companies and providing prediction services based on machine learning techniques, companies have valuable data from their customers, so they can act effectively and accurately on certain issues. However, the prediction of isolated client's lawsuits often does not reflect the robustness of a lawsuit prediction service. When it comes to electric power companies, the service provided encompasses a great number of customers at the same time (neighborhoods, cities, states). For example, the Equatorial Energy Group is responsible for the distribution of electricity in the states of Para, Maranhão, Piauı, and Alagoas, with over 5 million customers in Brazil.

Often, a customer's dissatisfaction with the service delivered can be motivating for other neighboring customers to file a lawsuit against the company. A practical example could be when a customer who has had the power supply interrupted for 24 hours is appointed as a potential customer to initiate a lawsuit through a prediction system. Usually, this interruption is not individual, so neighboring customers may also have this same profile and thus could potentially seek justice.

A tool for spatial analysis and spatially based prediction of consumer lawsuits against an electric power company that enables the company and its executives to look at the prediction of a group of customers in different or even risky locations provides a much more robust mechanism for addressing the problem. Because executives have such information, they can address more accurately the concerns of customers (or neighboring customer groups) at greater risk of filing lawsuits, thereby reducing costs to the company and, above all, increasing customer satisfaction and positive relationships with the company.

For the reasons mentioned above, the present work propose a tool for spatial analysis and spatially based prediction of consumer lawsuits against an electric power company using data mining and machine learning techniques. Because it is a tool for prediction of lawsuits, the paper presents a series of benefits, which we can highlighted as follows:

- A way of customer group prediction visualization and analysis for electric power companies.

- The system is applied industrially and with real data.
- Assists the company in establishing the risk of receiving legal proceedings.
- Helps managers and executives understand the motivation of lawsuits to generate planning and prevention actions.
- Helps improve customer service and response based on a robust and intelligent method.
- Helps improve client's satisfaction by avoiding lawsuits.

The rest of the paper is organized as follows. Section II presents the main related works. Section III presents the proposed tool of prediction of consumer lawsuits. The results are given in Section IV. Finally, a conclusion on the results obtained is given in Section V.

## II. RELATED WORK

Despite being techniques with huge social appeal and great concern among power companies, computational methodologies based on data mining and machine learning are still very scarce, especially for the purpose of providing a completely automatic method of predicting lawsuits. However, there is some work in the literature dealing with customer satisfaction, consumer profile analysis, and consumer churn in companies. These works will still be listed as related works because they were fundamental for the basis of the work proposed here, even if it is not possible to compare them directly with our proposed method.

Customers file a lawsuit because they want to be treated fairly by the company when a service failure occurs. Identifying customers who are likely to file a lawsuit against the company can also be considered as an identification of customer dissatisfaction. The company suffers a considerable monetary loss when some clients leave it, in addition to the procedural costs generated. This is classified as Customer Churn Prediction (CCP), where techniques based on machine learning, regression analysis, and predictive modeling are used to estimate the likelihood that customers will leave the company. CCP is a common problem in sectors such as telecommunications, [5], banking [6], e-commerce [7], and gaming [8].

A growing demand for new customers intensifies competition among commercial banks. To increase profits for ongoing operations and increase core competitiveness, commercial banks must avoid losing customers while at the same time acquiring new customers. In [6], the churn of commercial bank customers is predicted based on the Support Vector Machine (SVM) model and uses the random sampling method to improve the SVM model. When the ratio is 1:10 (churners:non-churners), the model has better results. Gordini and Veglio [7] developed a tailor-made churn prediction model for the SVM-based Business-to-Consumer (B2C) industry.

Amin et al. [9] propose a decision-making technique based on the Approximate Set Theory (AST) to extract important decision rules related to customer turnover. The AST classification based on genetic algorithms outperforms other rule generation algorithms used. Subsequently, the authors proposed a new CCP approach based on the concept of estimation of classifier certainty using distance as a factor [5]. It was found that the distance factor is strongly related to the certainty of the classifier.

Milosevic et al. [8] also presented a methodology for preventing customer churn, however, focusing on the game industry freemium. They evaluated learning models, such as decision trees, logistic regression, random forest, gradient boosting, and naive Bayes for churn prediction. The gradient boosting model showed better performance than the others. They also proposed a personalized churn prevention approach, identifying game features that are potentially interesting to the user and using them to customize notifications.

The relationship between previous events, which could be handled individually, with the quality and fidelity of the relationship is the purpose of Francisco et al. [10] proposed work involving a mobile phone company. The study confirmed (directly and indirectly) the assumptions of a positive and significant effect between satisfaction, trust and commitment and their antecedents.

According to Siu et al. [11], customers complain because they want to be treated fairly by the company when a service failure occurs. This study investigates the role of justice in retaining customers who had failed restaurant service experiences. As a result, the authors confirm the relationship of fairness between prior and subsequent satisfaction.

In a previous study, we proposed a method of predicting unregistered power consumption lawsuits and related variables [12]. The method proved to be robust in the task of classifying these types of customers, however, in our first study, we attempted to build an effective computational method and not a tool for corporate use. Also, the method in our previous work was not an extensible method, using only eXtreme Gradient Boosting. Thus, in this work, we propose the development of a tool for spatially based prediction of consumer lawsuits against electric power companies in a way that is intuitive and helps companies' managers increase customer satisfaction. This paper differs from our first work in several points, mainly in the fact that our present work is an extensible method which allows the insertion of new classifiers and offers a usable tool. Also, our previous work did not use the entire database, only a small proportion for the elaboration of the method. In the current work, we use the entire database of an electricity company.

Based on what is observed in the literature, and the importance of preventing customer lawsuits, the method proposed in this paper is based on the construction of a predictive model. For this, we use features extracted from the temporal relationship data of consumers of an electric company. The goal is to detect, in advance, cases where the customer may be dissatisfied and may go to court against the power company, and thus provide preventive information for managers and technicians to best address the problem.

We can see that many works propose ways to predict possible customer complaints, intention to leave the company or prosecute it. The works above mentioned do not offer a tool for individual or joint customer analysis. It is observed that the joint relationship of customer groups is something to be explored since features of customer groups in certain regions can be crucial in deciding lawsuits, especially in the case of customers of electric power services. The dissatisfaction of a customer can motivate his neighbors to file lawsuits against the company.

This paper presents a work done to provide electrical power companies and their executives with a tool for spatial analysis and spatially based prediction of consumer lawsuits. We propose a prediction method based on the customer behavior within the company; these predictions are plotted and visualized spatially. Thus, the company executives have a precise view of the problem and may take action to reduce customer dissatisfaction.

## III. MATERIALS AND PROPOSED METHOD

The methodology is organized into the following five steps: data acquisition, feature extraction, training, prediction, analysis and spatial visualization. These steps are described in the next subsections.

### *A. Data acquisition*

This work uses a private database from Equatorial Maranhão Energy Distributor. The database has customer information from various company sectors. Such information involves consumption history, supply discontinuity, financial information, customer complaints, equipment used for each one, etc.

The database used is accompanied by the ground truth for each customer, that is used in the prediction step to calculate validation metrics. Some customers already had a lawsuit against the company. Thus, it is possible to build a method to learn these patterns and later apply the pattern model to new customers.

After the data acquisition step, the data is organized into semantic groups to identify which features are relevant for the prediction of lawsuits. Also, we consider the generation of new features from these. This is done as good predictions require representative features for the classifier.

### *B. Fature extraction*

In this step, we analyze what features are considered relevant to build the electric company customer profile. Also, we describe the techniques used to refine and create new representative features. The features that we used are:

- General Information: individual features information of each client like spatial location, neighborhood, type of client (residential or commercial).
- Power consumption: features about consumption profile of each client.
- Power Loss: features related to occurrences of loss of energy or failure of distribution.
- Invoices: features that show if there were invoices caused by having UPC with that customer.
- Financial: historical payment behavior of a particular customer.
- Law: previous legal actions taken by the customer.

All features are preprocessed before the recognition step. In addition, it is necessary to analyze customers with a temporal perspective, that is, from customer information up to a certain period and predict their possible behavior for the later period. Thus, in addition to making data processing necessary, it is relevant to simulate time intervals in the database. To do this, we modeled the techniques for handling features with Feature Tools (an open-source Python tool for automated feature engineering), as described below:

- Numeric features are normalized, i.e. their values are converted to a single scale, that is, a single range of values;
- Categorical features are converted to numeric, using one hot encoding [13], which creates binary variables for each category;
- Mutual features are created using descriptive statistics: mean, median, standard deviation. Moreover, such mechanisms are used because of a temporal perspective of the information, for example, the standard deviation of consumption of each customer is calculated taking into account their history in the last 18 months.

Thus, the data were processed to aggregate temporal information using statistics. New features were created from the initial information, enabling the analysis of customer behavior at various time intervals. Below, we present the procedure of training of classifiers of the methodology.

### *C. Training*

The training step is responsible for choosing the best algorithm to recognize, based of previously computed features, if the client has intentions to file a lawsuit. We evaluate two algorithms: Extreme Gradient Boosting and Balanced Bagging Classifier. To evaluate, it is important to define metrics. In this paper, the chosen metric was sensitivity, since, in the case of lawsuits, the important thing is to be able to predict the largest number of possible lawsuits. The following are the two classifiers evaluated so far by the method:

- Extreme Gradient Boosting [14]: Tree-based machine learning algorithm that implements an optimization of the Gradient Boosting method, which is a technique that produces a predictive model from the union of less robust classifiers, showing better performance than if they were used in isolation.
- Balanced Bagging Classifier [15]: It consists in an unbalanced data handling technique, which selects by default 10 subsets of the initial data and performs random sampling.

The large number of individuals present in the database makes it impossible to train multiple models due to time and memory limitations. Thus, we divide the database randomly into 60% training and 40% for testing. From training, 10% of the data is taken to validate the model. Sampling is made keeping all individuals who filed a lawsuit and three times more individuals who did not file a lawsuit (1:3). The test dataset is used in the next step and remains as it is. All models are submitted to the same database and ranked according to the specified metric.

In the training step, the best classifier is chosen. As already mentioned, two classifiers were evaluated in this step. These are Extreme Gradient Boosting and Balanced Bagging Classifier. During training, each model generated by these classifiers is validated, and the classifier parameters are estimated using Random Search [16]. After a series of iterations, the best result achieved in validation is maintained as the best estimated classifier.

Once this is done, the classifier is submitted to the next step of the method, that is the prediction on a test dataset, as described below.

### D. Prediction

After the training step, with the best classifier selected, the best model is applied to the rest of the database. Thus, for each customer of the company, it is possible to measure the percentage probability that the customer will initiate a lawsuit or not. A client who is going to initiate a UPC lawsuit is considered to be one that has more than 50% probability, according to the previously estimated model.

To validate the prediction results of the model as a whole, global metrics are calculated in the test dataset to verify the robustness of the generated model. For this, metrics of sensitivity, specificity, and accuracy are extracted.

Accuracy (Acc) represents the proportion of true results (true positives and true negatives) in the population.

$$Acc = \frac{TP + TN}{TP + TN + FP + FN} \quad (1)$$

Sensitivity (Sen) expresses the number of true positives (clients with UPC correctly classified) divided by the total number of positive cases.

$$Sen = \frac{TP}{TP + FN} \quad (2)$$

Specificity (Spec) represents the number of true negatives (customers without UPC correctly classified) divided by the total number of negative cases.

$$Spec = \frac{TN}{TN + FP} \quad (3)$$

After all predictions of all consumers are generated and stored, we propose a tool for analysis and visualization of these data. Based on the latitude and the longitude information of each client, it is possible to show them on the map, and, with the prediction values of each one, it is possible to make a spatial analysis of the lawsuits.

### E. Spatial analysis and visualization

As a central point in the prediction of lawsuits, we have the identification of the clients with high probability to file a lawsuit and the determination of the more important causes that lead the predictor to this classification.

The Tobler's first law of geography states that "everything is related to everything else, but near things are more related than distant things". Based on this, we have that spatially grouped clients in general receive the same service quality and experience the same troubles in the consumer/provider relationship. In this work we used a geographic information system to manage and visualize several aspects of the lawsuits prediction, thus improving the company capacity to understand the problems and launch action plans to avoid the consumer dissatisfaction.

We theorize that the geovisualization of lawsuit predictions and their variables that strongly influence this prediction may offer an understanding of the dissatisfaction cause and the need actions the must be performed to increase the service quality. In this sense, we propose and develop a visualization tool that offers a visualization of prediction evaluation in different ways. This tool is shown in Figure 1.

Equatorial Maranhão's consumers are defined by contract accounts, and several users are linked to a traffic, so in the analysis, users of the tool can search for isolated consumers and/or traffic (Figure 1(A)). As an end-user visualization and analysis tool, it has a number of mechanisms that make it easy to search processes through filters (Figure 1(B)). Because it comprises an entire state of Brazil, there is a filtering option by Region (North, South, East, Northwest, Center); when selecting the region, the sectional and city areas are defined, which the user can also select through the filters; type of fare used; and the number of consumers that will be displayed (5000, 10000, or 50000). Because the prediction presents the probability by consumers, the user is free to define the percentage of probability they want to see on the map (Figure 1(C)). Furthermore, it is known that several features are extracted in the training step, and the prediction is evaluated by the features that influence the most the classifier decision, i.e., the variables that strongly influence this prediction. Therefore, the tool allows the visualization of the 10 most important features for each client and/or group of clients (Figure 1(D)). By clicking on the clusters (green, yellow and blue circles with number of consumers), this tab is redefined with the features of that customer group.

Another analysis that can be done is to select the client individually (blue marker). When this is done, a new field displays the contract accounts, the total probability, and the top ten features associated with the likelihood of triggering a lawsuit (Figure 2). Finally, the visualization and analysis is dispersed on the map defined by clusters or individual consumers, from the selection of filters. Through this series of mechanisms, managers are able to act accurately on the problem, trying to remedy problems with lawsuits and consequently reduce customer dissatisfaction.

In the next section, we will discuss the results of the proposed method steps and how the use of a geoanalysis and visualization tool can be crucial in basic service delivery companies such as Equatorial Maranhão.

## IV. Results and Discussion

This section presents and discusses the results obtained with the proposed method for the prediction of power consumption lawsuits. For evaluation, the acquired database was divided into two sets: training and test.

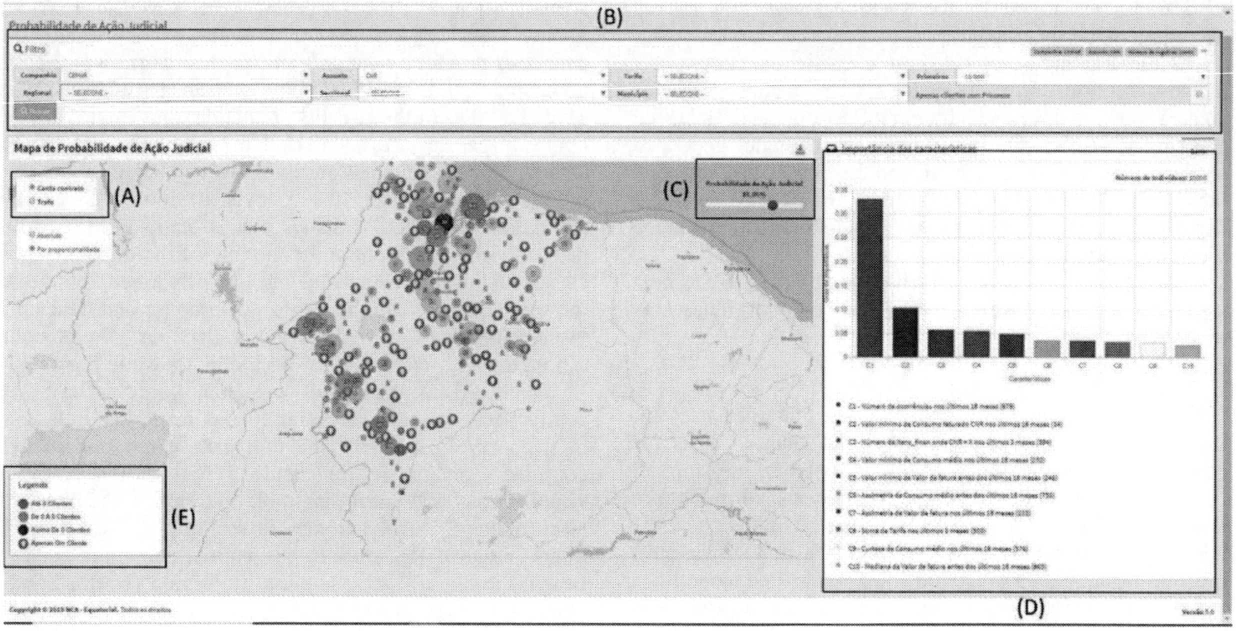


Figure 1. Spatial analysis and visualization tool for customer lawsuits prediction.

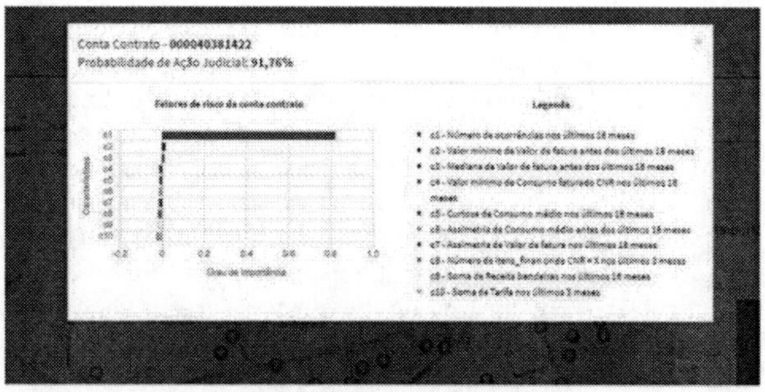


Figure 2. Individual analysis by contract account.

The client can go to court for a number of issues, which are defined during the court proceedings. In this paper the subject approached for prediction of lawsuit was Unregistered Power Consumption (UPC). This matter was chosen because it is the subject with the most lawsuits and that generates the most expenses for Equatorial Maranhão. The distribution of the proportion of clients with and without UPC is shown in Table I.

TABLE I. THE PROPORTION OF CLIENTS WITH UNREGISTERED CONSUMPTION SUBJECT IN TRAINING AND TEST DATASET.

| Dataset | Consumer with UPC | Consumer without UPC | Total |
|---|---|---|---|
| Train | 8.560 | 1.476.042 | 1.484.602 |
| Test | 5.714 | 998.442 | 1.004.156 |
| Total | 14.274 | 2.474.484 | 2.488.758 |

Approximately 2.5 million consumers were analyzed with the proposed method, however, a large imbalance in the datasets is noticeable. In the feature extraction step (Subsection III-B), 245 variables were retrieved directly from the database, while 925 were created from these by their temporal analysis, the transformation of categorical variables with one hot encoding and the use of descriptive statistics, totaling 1170 features.

Then, in the training step (Section III-C), the best classifier was decided when evaluating the 10% reserved for validation. Extreme Gradient Boosting yielded 90.3% sensitivity results in this step, while Balanced Bagging reached 93.6%. For this reason, Balanced Bagging was chosen to be the classifier that will make the prediction in the next step.

At the prediction step, the remainder of the database was evaluated by the model created in the training step, and thus generated the likelihood of a client to go to court against the company. To assess the robustness of the method, global validation metrics were generated, which are shown in Table II.

TABLE II. RESULT OF PREDICTION STEP USING BALANCED BAGGING.

| Subject | Acc (%) | Sen (%) | Spec (%) |
|---|---|---|---|
| UPC | 91.86 | 96.52 | 91.83 |

As can be seen from Table II, the Balanced Bagging classifier proved robust results in the classification of new customers. Company managers can use this information, which has 96.52% sensitivity, in predicting UPC to avoid legal costs and improve quality.

However, despite this important information, there is a need to present the information more accurately, not just numerically. So, we use spatial analysis to visualize more effectively large or micro relationships between clients. As presented in Section III-E, a number of filtering mechanisms are presented for better map viewing. These mechanisms are crucial for geoanalysis, as executives can go directly to high-risk regions that are most likely to start legal action.

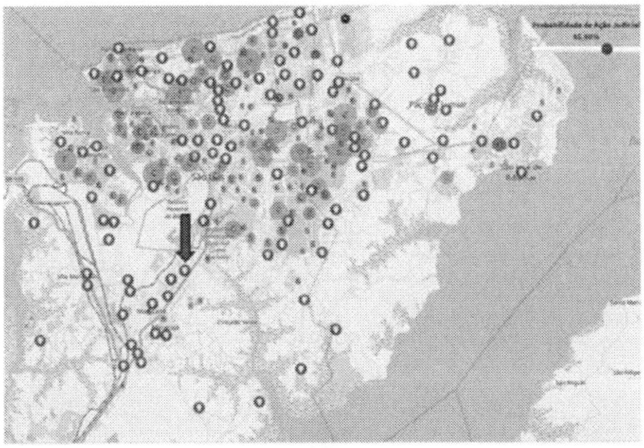

(A)

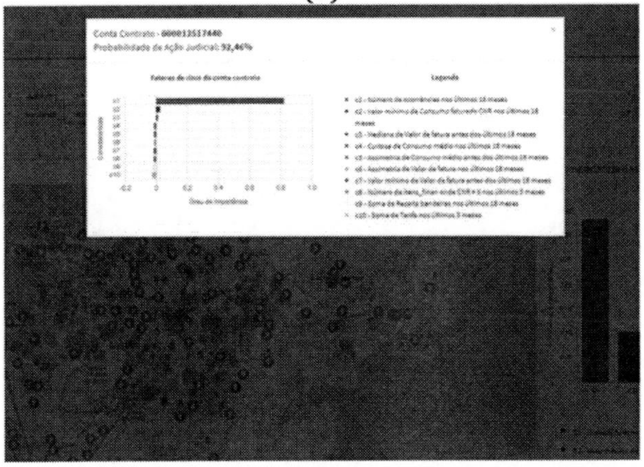

(B)

Figure 3. Case study: (A) Selecting an individual consumer, blue marker and (B) Display of variables that influence that consumer to file a lawsuit.

Based on an individual analysis, the manager can see which variables will influence that customer to start legal action against the company and act incisively. An example of this analysis in a case study can be seen in Figure 3 where, after filtering the interface, the user selects the individual consumer (blue marker, Figure 3A) and a new field with information of the most influential variables is presented (Figure 3B). By analyzing the variables, it is possible to obtain what influences the client the most, for example, the lack of energy in recent days. We can note that the customer has certain features that can influence him to file a lawsuit against the company, and that there is a 92.46% probability of going to court (Figure 3B). Also, the most probable cause is, for example, power outages. Then, the tool user can select the customer group (green, yellow and red circles), which that individual customer is contained in and the variables will be updated. So, managers will realize that not only a single customer, but the neighborhood is also going through the same problem and that the company's performance needs to reach that group of consumers to increase customer satisfaction with the company. Thus, the tool not only shows the percentage of dissatisfaction generated by the classifier, but also presents the possible variables that influenced consumers and groups of consumers to trigger actions, all combined with a friendly and intuitive geovisualization and analysis tool.

It is important to note that this is just a case study in which the tool can facilitate the use of company executives. Because it is a basic service delivery company and encompasses an entire state of the federation, the use of computational tools to help improve satisfaction is increasingly needed. A spatial analysis tool that aggregates lawsuit prediction information, as well as facilitating the company's operations, also proves to be an ally in the constant search for improvements in customer service.

## V. CONCLUSION

This paper presents a spatial geoanalysis tool which predicts an electric power company customer's behavior. The prediction system is designed to allow its results to be used to avoid public disputes, e.g., in court trials. The work presents a prediction system based on computational intelligence techniques for prediction in a company from Maranhão state, Brazil. The method was applied to a database of more than two million customers and proved to be robust in hitting customers coming with UPC lawsuits against the company, resulting in a sensitivity of 96.52%.

Also, the geoanalysis tool presents the predictions of each client along with the variables that most influenced each of them. This tool has a user-friendly interface and several features for customer and customer group analysis, allowing company managers to act incisively on issues in a variety of areas that the company understands by reducing legal costs and increasing customer satisfaction. The tool also allows a complete visualization and analysis of the most diverse areas and the most diverse consumers of the state.

However, improvements are suggested as future work, such as acting on new judicial subjects (not only in UPC), using more classifiers in the training and prediction step, and generating other forms of data visualization such as heatmaps. Other spatial features can be added to search for improvements, analyze the various features of the database and verify the importance of spatial information for the robustness of the method, implement a tool in other basic service companies and, finally, create new features that can improve the classifier.

## ACKNOWLEDGMENTS

The authors would like to thank Equatorial Energy for the financial support provided through the National Electric Energy Agency (ANEEL) Research and Development Program (R&D), PD-00037-0031/2019.

## REFERENCES

[1] P.-N. Tan, Introduction to data mining. Pearson Education India, 2018.

[2] M. Mohri, A. Rostamizadeh, and A. Talwalkar, Foundations of machine learning. MIT press, 2018.

[3] V. A. Ibanez, P. Hartmann, and P. Z. Calvo, "Antecedents of customer loyalty in residential energy markets: Service quality, satisfaction, trust and switching costs," The Service Industries Journal, vol. 26, no. 6, 2006, pp. 633–650.

[4] R. Johnston and A. Fern, "Service recovery strategies for single and double deviation scenarios," Service Industries Journal, vol. 19, no. 2, 1999, pp. 69–82.

[5] A. Amin, F. Al-Obeidat, B. Shah, A. Adnan, J. Loo, and S. Anwar, "Customer churn prediction in telecommunication industry using data certainty," Journal of Business Research, vol. 94, 2019, pp. 290 – 301.

[6] B. He, Y. Shi, Q. Wan, and X. Zhao, "Prediction of customer attrition of commercial banks based on svm model," Procedia Computer Science, vol. 31, 2014, pp. 423 – 430, 2nd International Conference on Information Technology and Quantitative Management, ITQM 2014.

[7] N. Gordini and V. Veglio, "Customers churn prediction and marketing retention strategies. an application of support vector machines based on the auc parameter-selection technique in b2b e-commerce industry," Industrial Marketing Management, vol. 62, 2017, pp. 100–107.

[8] M. Milosevic, N. Zivic, and I. Andjelkovic, "Early churn prediction with personalized targeting in mobile social games," Expert Systems with Applications, vol. 83, 2017, pp. 326 – 332.

[9] A. Amin et al., "Customer churn prediction in the telecommunication sector using a rough set approach," Neurocomputing, vol. 237, 2017, pp. 242–254.

[10] E. C. Francisco-Maffezzolli, P. H. M. Prado, W. V. da Silva, and R. Z. Marchetti, "Evaluation of the customer relationship quality and propensity to change mobile telephone operators," Brazilian Business Review, vol. 8, no. 4, 2011, pp. 1–22.

[11] N. Y.-M. Siu, T. J.-F. Zhang, and C.-Y. J. Yau, "The roles of justice and customer satisfaction in customer retention: A lesson from service recovery," Journal of business ethics, vol. 114, no. 4, 2013, pp. 675–686.

[12] F. Y. Oliveira et al., "Prediction of unregistered power consumption lawsuits and its correlated factors based on customer data using extreme gradient boosting model," in 2019 IEEE International Conference on Systems, Man and Cybernetics (SMC). IEEE, 2019, pp. 2059–2064.

[13] J. Buckman, A. Roy, C. Raffel, and I. Goodfellow, "Thermometer encoding: One hot way to resist adversarial examples," in International Conference on Learning Representations, 2018.

[14] T. Chen and C. Guestrin, "Xgboost: A scalable tree boosting system," in Proceedings of the 22nd ACM sigkdd international conference on knowledge discovery and data mining. ACM, 2016, pp. 785–794.

[15] G. Lemaıtre, F. Nogueira, and C. K. Aridas, "Imbalanced-learn: A python toolbox to tackle the curse of imbalanced datasets in machine learning," The Journal of Machine Learning Research, vol. 18, no. 1, 2017, pp. 559–563.

[16] J. Bergstra and Y. Bengio, "Random search for hyper-parameter optimization," Journal of machine learning research, vol. 13, no. Feb. 2012, pp. 281–305.

# Spatio-Temporal Analysis of Premature Mortality Trends in the United States

Yelena Ogneva-Himmelberger
Department of International Development, Community and Environment
Clark University
Worcester, MA, USA
e-mail: yogneva@clarku.edu

*Abstract*— **This paper applies geospatial data analytics to explore trends in premature mortality in the United States. Premature mortality, or Years of Potential Life Lost (YPLL), is one of the public health measures that focuses on deaths that happened at younger ages (before the age of 75 years) and thus could have been prevented. We used publicly available YPLL data for 2005-2016 for 3080 United States counties with spatio-temporal data mining tools in Geographic Information Systems (GIS) software to create a space-time cube and to find temporal trends in this measure. Our preliminary results indicate that 22% of counties experienced a statistically significant upward trend in YPLL, 24% experienced a statistically significant downward trend, and the remaining 54% did not experience any monotonic trend in YPLL measure. These findings can help county-level department of public health with developing targeted interventions to reverse upward trends in YPLL measure.**

*Keywords-GIS; Spatial and spatio-temporal statistics; trend analysis; health; premature mortality.*

## I. Introduction

Premature mortality, or Years of Potential Life Lost (YPLL) is one of the public health measures that focuses on deaths that happened at younger ages (in the United States, before the age of 75 years) and thus could have been prevented. The Center for Disease Control collects annual county-level mortality data and provides state-level analysis. However, more detailed, local-level analysis is needed in order to conduct surveillance of temporal trends in premature mortality, and to evaluate the effectiveness of program interventions [1]. The goal of this paper is to identify temporal trends in premature mortality in the United States, using county-level data and space-time techniques in Geographic Information Systems (GIS). The rest of the paper is organized as follows: Section II provides a description of our data sources and methods. Section III reports our preliminary results, and Section IV discusses future research steps.

## II. Data and Methods

The YPLL measure was obtained from the County Health Rankings website [2] for 2005-2016. Since premature deaths are relatively rare events, YPLL is based on a three-year period, rather than on a single year. In this dataset, YPLL is a rate per 100,000 people and is age- adjusted to the 2000 US population. The rate is calculated as the number of total years of potential life lost for deaths that occurred amongst people who reside in a county under age 75, divided by the aggregate population under age 75 for the three years. The number of years lost is calculated for each death individually, and is based on the age at the time of death. The younger the person, the higher the number of years lost. To map the YPLL data, the GIS layer of county boundaries was downloaded from the U.S. Census Bureau [3] and YPLL tables were joined to the GIS layer using county Federal Information Processing Standards codes. There are 3142 counties in the United States, but 62 of them were missing YPLL data.

To identify spatio-temporal trends in YPLL, first, a space-time cube was created in ArcGIS Pro [4]. A space-time cube is a collection of spatial units (in this case, counties) layered vertically according to time. The bottom layer of the cube corresponds to 2005, the earliest year in the dataset, and the top layer of the cube corresponds to 2016, the latest year. Thus, a particular county at a given year is referred to as a bin within the space-time cube.

We applied the Mann-Kendall trend test to identify statistically significant temporal trends in YPLL for each spatial bin. This test compares values within each spatial bin over time and calculates changes between each consecutive time steps [5]. It identifies consistently increasing or decreasing trends over time at a location. Previous research applied this technique to analyze temporal trends in ground water level and precipitation [6]-[8], aridity [9], air temperature [10], traffic accidents [11] and crime [12]. We apply this widely used technique in a new context related to human health.

The output from the Mann-Kendall test is a spatial layer showing each spatial bin belonging to one of the seven categories: Up trend – 99% confidence; Up trend – 95% confidence; Up trend –90% confidence; Down trend – 99% confidence; Down trend – 95% confidence; Down trend – 90% confidence; No significant trend. The confidence level is determined based on the z score and the p-value of the trend.

To visualize the trends, we mapped seven trend categories and calculated the number of counties within each category.

## III. Results

To create the space-time cube, we first created a separate YPLL map for each year. An example of such map is provided in Figure 1.

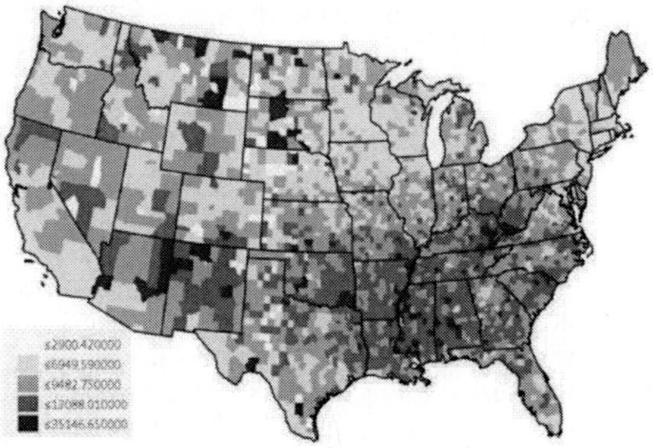

Figure 1. Premature mortality rate in continental United States in 2016.

Then, we combined all eleven maps into a space-time cube. It consisted of 3080 locations (counties) and 11 time slices, resulting in 33880 space-time bins. The minimum, mean, and maximum values of YPLL in the cube were 2817, 7963 and 35147 years/100,000 people, respectively. The Mann-Kendall test results are shown in Table 1 and Figure 2.

TABLE 1. NUMBER OF COUNTIES IN EACH TREND CATEGORY

| Trend | Number of counties |
|---|---|
| decreasing - 99% confidence | 346 |
| decreasing - 95% confidence | 258 |
| decreasing - 90% confidence | 149 |
| no trend | 1643 |
| increasing - 90% confidence | 157 |
| increasing - 95% confidence | 249 |
| increasing - 99% confidence | 278 |

These results indicate that 684 (22%) counties experienced a statistically significant upward trend in YPLL, 753 (24%) counties experienced a statistically significant downward trend, and the remaining 1643 (54%) counties did not experience any monotonic trend in YPLL measure. These findings are important, as they highlight areas with the alarming trend (statistically significant increase in YPLL rates over time) and can help local departments of public health develop targeted interventions to reverse these trends.

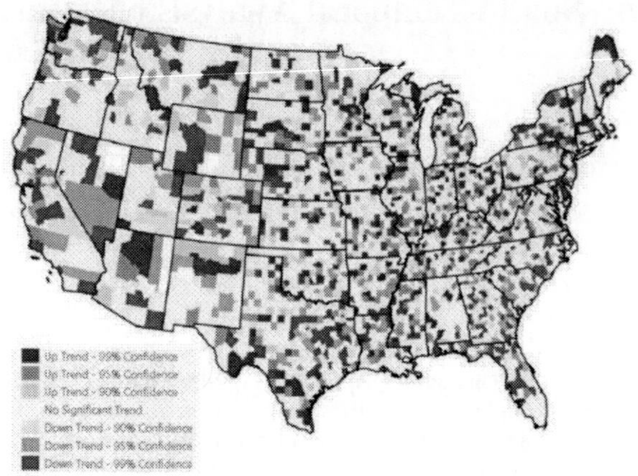


Figure 2. Mann-Kendall trends in premature mortality rates in continental United States, 2005-2016.

## IV. Conclusion

In this paper, we identified temporal trends in premature mortality in the United States, using county-level data and space-time techniques in Geographic Information Systems. Our preliminary results indicate that in each state, there are counties with both upward and downward YPLL trends, sometimes next to each other. As the next step, it would be important to select several states for an in-depth analysis of the relationship between socio-economic and demographic factors and health outcomes to gain a better understanding of factors related to local variations in YPLL rate. This work could provide additional insights for more effective public health interventions.

## References


[1] Centers for Disease Control and Prevention. "Premature mortality in the United States: Public health issues in the use of years of potential life lost. MMWR Morb Mortal Wkly Rep. 1986;35(suppl 2), pp.1S-11S.

[2] County Health Rankings. Available from: https://www.countyhealthrankings.org/ [retreived: February, 2020]

[3] U.S. Census Bureau. Available from: https://www.census.gov/geo/maps-data/data/cbf/cbf_counties.html [retreived: February, 2020]

[4] ESRI, 2019. ArcGIS Pro Help. Available from: https://pro.arcgis.com/en/pro-app/help/main/welcome-to-the-arcgis-pro-app-help.htm [retreived: February, 2020]

[5] M. G. Kendall, Rank correlation methods, 2 ed. Oxford, England: Hafner Publishing Co. 1955.

[6] R. Agarwal and P. K. Garg, "Statistical assessment of groundwater resources and long term trend using geospatial techniques," 2016 IEEE International Geoscience and Remote Sensing Symposium (IGARSS), 2016, pp. 1808-1811.

[7] A. Chandrakar, D. Khare, and R. Krishan, "Assessment of spatial and temporal trends of long term precipitation over

Kharun watershed, Chhattisgarh, India," Environmental Processes, vol. 4, no. 4, pp. 959-974, December 2017.

[8] S. Kumar, D. Machiwal, and D. Dayal, "Spatial modelling of rainfall trends using satellite datasets and geographic information system," Hydrological Sciences Journal, vol. 62, no. 10, pp. 1636-1653, July 2017.

[9] F. J. Moral, L. L. Paniagua, F. J. Rebollo, and A. García-Martín, "Spatial analysis of the annual and seasonal aridity trends in Extremadura, southwestern Spain," Theoretical and Applied Climatology, vol. 130, no. 3, pp. 917-932, November 2017.

[10] F. Viola, L. Liuzzo, L. V. Noto, F. Lo Conti, and G. La Loggia, "Spatial distribution of temperature trends in Sicily," International Journal of Climatology, vol. 34, no. 1, pp. 1-17, 2014.

[11] M. K. Rahman, T. Crawford, and T. W. Schmidlin, "Spatio-temporal analysis of road traffic accident fatality in Bangladesh integrating newspaper accounts and gridded population data," GeoJournal, July 2017.

[12] L. Ross, Y. Ogneva-Himmelberger, and C. Starr, "The use of geographic information systems for real-time monitoring of comprehensive community initiatives," Justice Research and Policy, April 2019; DOI: 10.1177/1525107119843259

# A Microservices Approach for Parallel Applications Design: A Case Study for CFD Simulation in Geoscience Domain

Alexey Cheptsov
High Performance Computing Center Stuttgart,
University of Stuttgart
Stuttgart, Germany
e-mail: cheptsov@hlrs.de

Oleg Beljaev
Donetsk National Technical University
Computer Science Department
Pokrowsk, Ukraine
e-mail: oleg69@ukr.net

*Abstract*—**Current geoscience applications face two major challenges – the integration with numerous diverse sensor devices and the use in real-time use case scenarios. Whilst the challenge of service integration is addressed by the concept of Cyber-Physical systems, which aims to incorporate sensor data in application workflows, the usage of High Performance Computers helps minimize the execution time to fulfill the real time scenarios requirements. However, the existing programming models do not allow scientific workflows to take advantage of both technologies simultaneously. This paper contribution offers an approach to encapsulation of workflow-based applications into services, which are flexible enough to run on heterogeneous, distributed infrastructures spanning over both industrial sensor services and parallel computing systems. The approach is demonstrated on a computational fluid dynamics simulation study of aerodynamic processes in complex underground mine ventilation networks.**

***Keywords-Dynamic Simulation; Computational Fluid Dynamics; Microservices Architecture; Workflows; ChEESE.***

## I. INTRODUCTION

Geoscience applications rely largely on simulation, which is used to retrieve and investigate the state of the targeted complex dynamic systems and also to predict their behavior under certain conditions in the future. One of the typical geoscience simulation tasks is served by Computational Fluid Dynamics (CFD) – a technique that is used to study the behavior of liquids and gases in complex environments. The CFD technique can be used to model many safety-critical processes, such as, for example, the propagation of waves as a result of tsunamis, the distribution of volcanic plumes after an eruption, or the distribution of air and hazardous gases in underground ventilation objects like coalmines. As all the other CFD applications, these studies are based on complex mathematical models (like Navier-Stokes equation), which generally create a good deal of uncertainty for the simulation results and also require computationally expensive solution methods.

In practice, geoscience applications are often organized in workflows with several interconnected components, each implementing a specific part of the application logic and running on a dedicated resource of the distributed system. The computationally intensive parts of the workflow are usually executed on parallel High Performance Computing (HPC) resources, whilst the data acquisition happens on the sensor nodes. However, the workflow approach has several limitations. Firstly, the workflow management software requires quite a rich functionality of resource management, application scheduling, monitoring and other middleware that is related to the workflows execution (like Pegasus, as described by Chang et al. in [3]), which are difficult to provision on the production HPC systems. Secondly, the workflow-based specification of applications requires quite an extensive metadata schema, which might require substantial change from one execution scenario to the other. Lastly, the implementation of the control flow across the components that include parallelized applications, e.g. with the help of the Message-Passing Interface (MPI), is difficult due to the functional orientation of the latter. In other words, it is technologically difficult to build a workflow management system that would enable running applications of both types (event-based serial ones and functionally-oriented parallel ones) within the same control and data flow logic and on distributed heterogeneous resources.

This paper's contribution introduces an alternative approach, which allows MPI applications to be built in a service-oriented way, thus allowing for flexibility of data processing, as required by geoscience applications, whilst keeping a much lower management overhead than in the case of traditional workflow management systems. The proposed approach is facilitated by a Multiple Instruction Multiple Data (MIMD)-based programming model, which could be inheritably implemented in MPI-parallel applications. The use of the elaborated programming model is illustrated on an implementation of a CFD simulation application for underground coalmine ventilation tasks. The remainder of the paper is organized as follows: Section II gives an introduction of ventilation networks and simulation tasks for them. Section III elaborates a microservices based architecture and methodology for implementation of simulation applications. Section IV discusses the results that are obtained for the evaluation cases. Section V concludes the paper and discusses the main outcomes.

## II. VENTILATION NETWORKS AS OBJECTS OF MODELLING AND SIMULATION

Fossil coal remains one of the most important energy sources, along with gas, oil, and regenerative energy technologies. In particular, it holds a leading position among the fossil fuels with proven reserves of over 1 Tera-Ton (as shown in [4]) worldwide. At the same time, the coal industry is one of the most dangerous and security-critical industry branches, due to the complexity of the coal mining process from the deep underground locations (up to many tens of kilometers under the surface) and considering the aspects of a high gas content in the obtained coal masses. Ventilation is the most important aspect of the security provisioning in underground production areas of coalmines (see Figure 1) – it aims to ensure the underground mining workers with the necessary amount of fresh air and also to dilute the hazardous gases (mainly $CH_4$ – marsh gas) that are emitted during the coal loosing, transporting and other technological activities of the mining process.

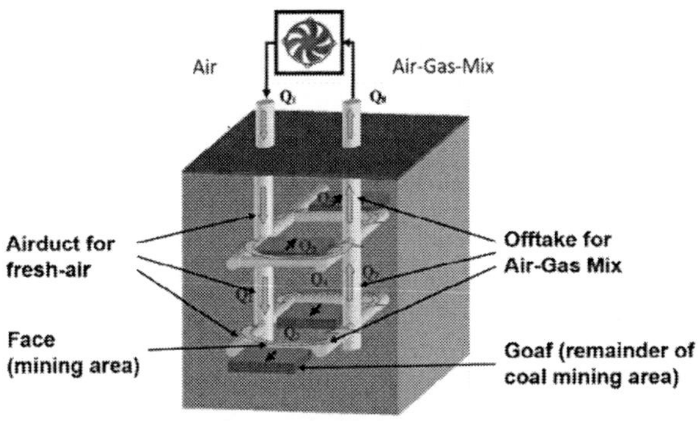


Figure 1. Structure of underground mine ventilation.

The degassing procedure is of especial importance for the safety of the mining process – the marsh gas concentration that exceeds the upper threshold can lead to vast exposures in the underground area with major human losses and injuries (see Figure 2). Historically, gas exposures have been the major reason of big catastrophes that have happened in the coal industry since the beginning of the industrialization era and up to nowadays (see [10]).

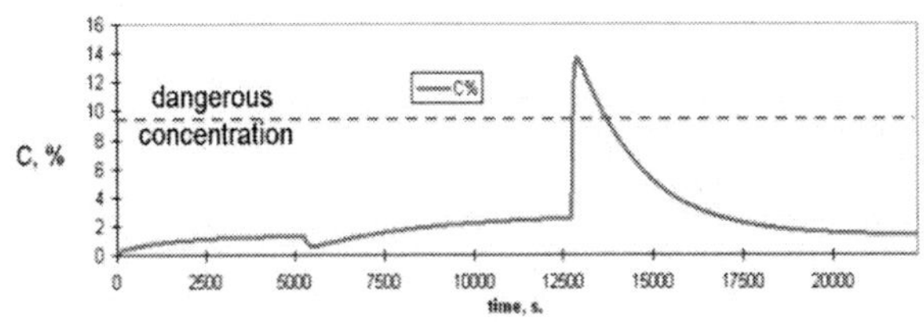


Figure 2. Analysis of march gas concentration in time.

The unpredictable nature of the gas emission is the major challenge for the operation of underground coalmines. Solving the challenge of air and gas distribution prediction requires a detailed knowledge of all dynamic processes that are happening in the elements of real industrial ventilation objects (see example in Figure 3). In most cases, the only possible chance to get insight into optimal planning of air distribution along the ventilation elements and plan the gas dilution actions is the use of modelling and simulation, coupled with the information coming from sensors, which are measuring airflow speed and gas concentration.

Typical CFD models of coalmine ventilation are based on the macroscopic definition of the multiphase flow based on Navier-Stokes equation, e.g., in the following general form for the air distribution in one branch of the ventilation network (e.g., as elaborated by Svjatnjy [5] for a 1-D approximation):

$$\begin{cases} -\dfrac{\partial P}{\partial \xi} = -\dfrac{2\rho}{F^2} Q \dfrac{\partial Q}{\partial \xi} + \dfrac{\rho}{F}\dfrac{\partial Q}{\partial t} + rQ^2 + r'(t)Q^2 \\ -\dfrac{\partial P}{\partial t} = \dfrac{\rho a^2}{F}\dfrac{\partial Q}{\partial \xi} - \dfrac{\rho a^2}{F} q \end{cases}, \quad (1)$$

where P is the pressure, Q is the airflow, t is the time, $\xi$ is the spatial coordinate, and the other values represent aerodynamic parameters and coefficients.

The gasflow distribution analysis is based on the data obtained from the sensors, which are fed in the transport equation (1) similarly to the airflow (for example, as described by Stewart et al. [6]). Given the insufficient coverage of the underground production areas by sensors, additional prediction models might be used, e.g., as described by the previous publication [7] for the goaf – a mine area that remains after the coal mining. The model of the whole ventilation network (see example in Figure 3) is built from the models of each of its elements/branches in the general form (1), extended by the boundary conditions in the nodes and following this hierarchical organization of the elements: approximation unit of numerical method → element/airway → section → network (see Figure 4). Further coalmine ventilation simulation tasks include reduction of the energy that is required for operation of the main mine fans (which usually consume up to 40% of the overall coalmine energy), as shown by Clausen [8].

The improvement of the quality of the CFD computational models has been the focus of many research activities in the last decade. In particular, lots of activities have been concentrated around integration of sensor data into the simulation process, for example, in the form of initial or boundary conditions for the mathematical models, serving the basis for the simulation packages. The "online" sensor data integration allows, among others, to specialize the generic models, i.e., adapt the model parameters to the specifics of the targeted simulation object. With the proliferation of the unified (Ethernet) networking standards (both wired and wireless) in the field of industrial and automated systems (like Industrial Ethernet's solutions PROFINET, Modbus, and others, as described by Kay et al. [1]), sensors have become a vital part of the distributed computing infrastructure and, most essentially, of their software applications. Such an infrastructure, which allows a seamless integration of the network-enabled data acquisition devices (such a CH4 – marsh gas sensor) with the "traditional" computation and storage facilities, is often referred to in the literature as Cyber-Physical Systems (CPS). Being initially elaborated for the automotive and industrial automation domains (as described by Broy [2]), the CPS concept is gaining an increasingly growing popularity for the other industrial and scientific areas, including the geoscience applications domain, as targeted by this paper.

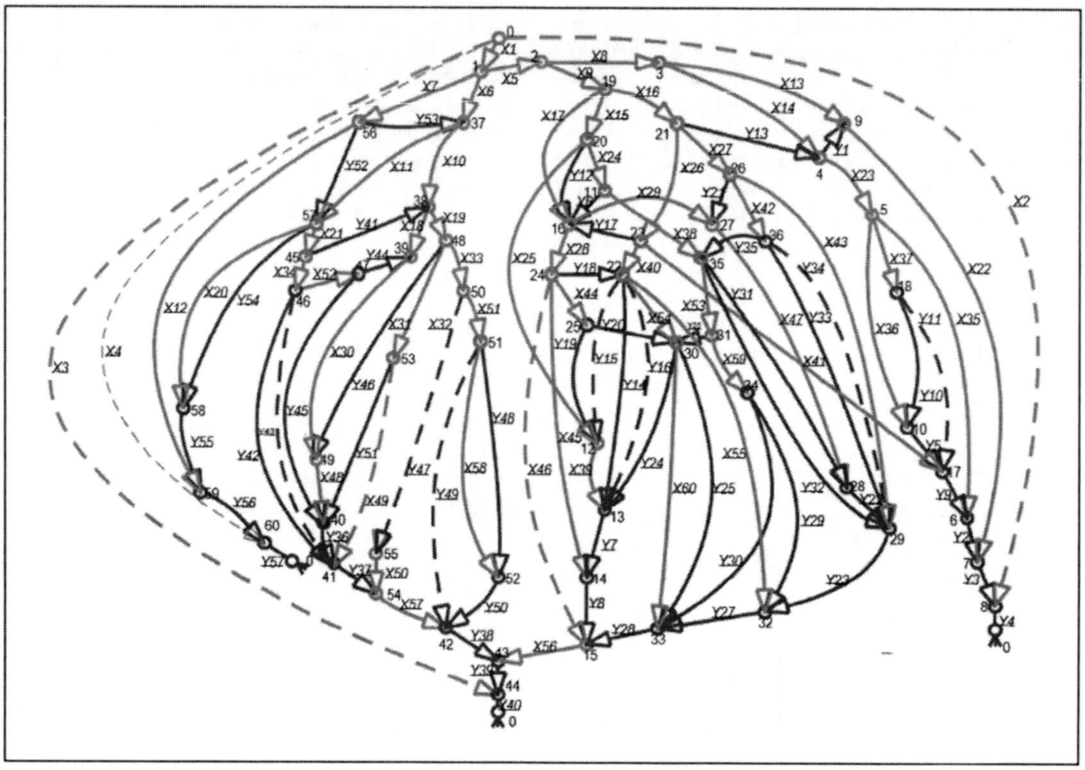

Figure 3. Illustration of real-complexity ventilation network with 117 branches and 61 connection nodes (coalmine South-Donbass Nr. 3 in Ukraine). The coloring of airflows is only used for better readability.

Nowadays, geoscience applications are represented by parallel software packages that require large-scale computing and storage facilities of HPC and Cloud infrastructures. Those applications are usually developed by means of MPI or other parallelization standards and have a limited ability to incorporate data from external (distributed over the communication network) sensors and other acquisition devices due to the following limitations:

- Heterogeneity of the CPS distributed infrastructure – many sensor devices are provided on the basis of a host system, whose architecture might differ from the typical HPC, Cluster, or Cloud environment, but still requires a seamless integration within the distributed application workflows. However, the standard parallelization approaches require a uniform infrastructure with the compute nodes of the same hardware architecture and performance class.
- Limited flexibility of the mainstream parallelization approaches to support distributed application scenarios – many parallel applications rely on the Single Instruction Multiple Data (SIMD) technique, which mainly targets densely built compute systems like HPC. However, the applications that are running on the truly distributed infrastructures (HPC + Cloud + remote embedded systems) have to be developed according to the MIMD approach, in order to allow different functionalities to be executed on different types of systems.

## III. Microservice Architecture for Simulation Applications Development

Ventilation networks analysis is a challenging process – the simulation software developers often face problems, some of which are listed below:

- Nonlinearity of the base equation system, which causes the need of applying a special numerical method, e.g., the Finite Differences, Finite Volumes, Discontinuous Galerkin, etc.
- Complex topological organization of ventilation networks.
- Complex hierarchical structure of ventilation elements involving several levels of control and regulation.

Although the mainstream simulation packages like OpenFOAM or ANSYS-CFX offer a rich development functionality which is sufficient for the implementation of the ventilation models, there are still numerous adaptations and optimizations necessary, which are very difficult to implement with general-purpose simulation tools. For example, it would be difficult to integrate the model for diffusion and filtration processes between the airways and gas emission sources in the ventilation section. However, the major disadvantage that remains is the inability of their use in distributed, heterogeneous hardware architecture environments. Therefore, novel approaches are required for the implementation of portable, scalable, and efficient simulation software.

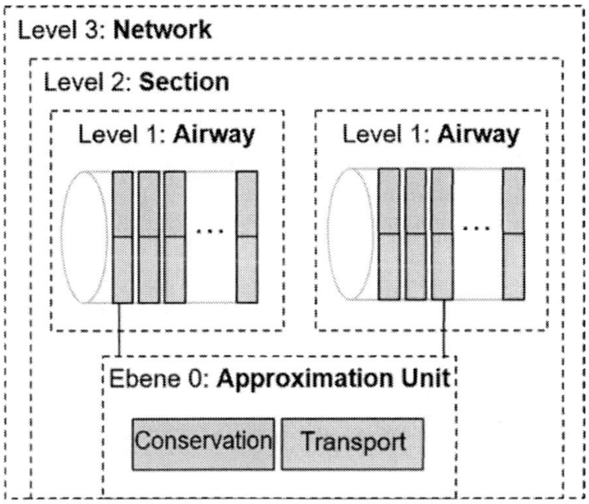


Figure 4. Hierarchical approach for composition of ventilation network models.

Object-oriented modelling and service-oriented platforms have been established in the last decade as an alternative to the traditional software development approaches. The actual trend in the service-oriented development goes in the direction of microservice (MS) architectures – a concept that is initially coming from the Internet-of-Things, Cyber-Physical Systems and Cloud domains. MSs are independent, isolated, portable software blocks/units, each implementing a part of a complex system, which can be decomposed according to functional, spatial or any other conditions. Each MS follows in its implementation the locality principle (as illustrated in Figure 5) – i.e., bearing responsibility for the assigned part of the complex system, according to the decomposition strategy. In order to reflect physical or informational connections of the real object, MSs can be bound together by a common data and/or control flow.

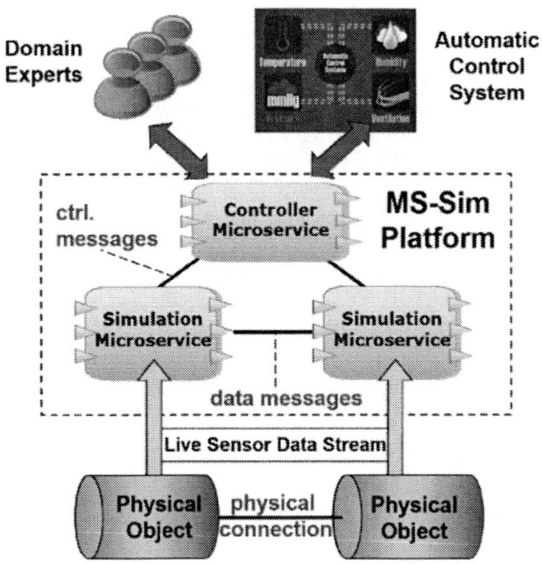


Figure 5. Microservice architecture for CFD simulation platform.

An example of a platform for execution of MS-based simulation workflows is presented in Figure 6 below. The connection between all the MSs in the system is provided by an external communication library, so that the MS-developers do not need to handle the data exchange explicitly – the data exchange performs asynchronously with the help of special buffers, used to flush the output data or read the input from the other MSs in the system, whenever required by the modelling algorithm. In case of an MPI-based implementation, every MS is executed by an independent MPI process, which is developed on the remote resource or a compute node. The MPI processes can run on a heterogeneous architecture, as they could be enabled by modern implementations like OpenMPI. Each simulation MS follows a command flow, as defined by the simulation logic, which can be implemented in an event-driven way, as shown by the listing in Figure 7. The command flow can be steered by a dedicated "master" MS, depicted as a "Controller" in Figure 5.

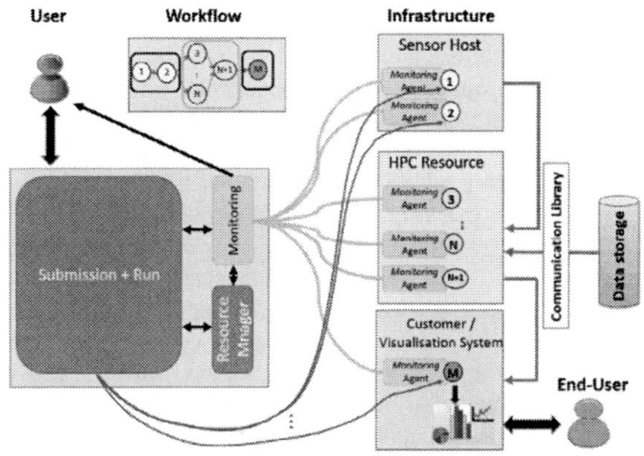


Figure 6. Execution platform design for microservice-based applications.

The communication between the MS happens with the help of the underlying communication library, e.g., by means of point-to-point or collective MPI calls to an MPI implementation. Results storing can happen either individually by every MS (e.g., in the Paraview format) or in the collective way using a consolidating database like ElasticSearch.

In fact, such functional decomposition-driven approach to the development of simulation software is not particularly new to the simulation of complex dynamic systems. For example, the Matlab/Simulink modelling package provides a module- (block-) based approach to construct a model from many smaller subsystems (functional blocks or submodels). However, this approach is inefficient when dealing with big, dynamic configurations of objects with a complex and variable network topology, like the targeted ventilation systems of coalmines. The main advantages of the MS-approach for application development are the separation of the computation and communication application logic during the development process, resulting in a decrease of implementation efforts, simplified implementation of the horizontally-scalable applications (just by replicating

```
void run() {
 bool finish = false;
 while (!finish) {
 // receiving command from the Master
   buffer_sync(Ports::command_flow);
   int command = get_buffer_value_stack<int>
                   (Ports::command_flow);
   switch (command) {
     case Commands::stop: {
       finish = true;
       stop();
       break;
     }
     case Commands::simulation: {
       simulation();
       break;
     }
     ...
 }
```

Figure 7. Example of workflow event handling by microservice.

The services), easy implementation of hierarchical relationships between the services of different functionality levels (horizontal scalability), and the possibility to deploy a MS on any resource of the heterogeneous infrastructure. In the work that is presented in the paper, a modular architecture for development of CFD simulation applications based on a library of MS-components has been implemented. A simulation application is logically organized as a modular assembly of various MSs within a common architecture according to the hierarchical composition of the elements/services, as was previously depicted in Figure 4. Moreover, using the modular approach, the MS-based simulation application can transparently perform experimentation with different approximation schemes, discretization approaches, numerical solution methods, results validation techniques, etc.

## IV. Implementation of CFD Ventilation Study with Microservices Architecture

The above-presented microservices approach was used to implement a simple CFD study for a ventilation section consisting of 2 airways ($Q_{FW}$, $Q_{VS}$), a coal mining area ($Q_S$), and a goaf ($q$) with a gas emission source ($q_m$), according to the topology depicted in Figure 8. The airflow was enforced by a single mine fan, connected to the first airway ($Q_{FW}$).

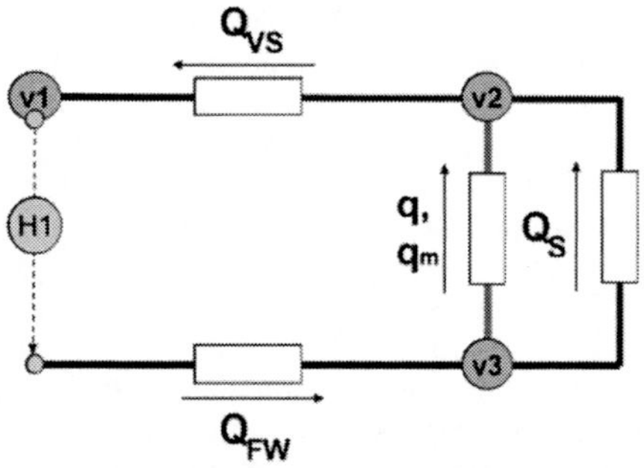


Figure 8. Test section structure.

Each element of the section is described by the base model (1). However, since the system (1) includes partial differential equations, the models development had to start from a lower granularity level – approximation elements of the numerical method of spatial discretization. In our case, the finite differences method was chosen for reasons of simplicity, that resulted in a set of k normal differential equations for approximation units in the general form (2):

$$\begin{cases} \dfrac{dQ_k}{dt} = \dfrac{F}{\rho}\cdot\dfrac{P_k - P_{k+1}}{\Delta y} - \dfrac{F}{\rho} r Q_k^2 - \dfrac{F}{\rho} r(t)\cdot Q_k^2 \\ \dfrac{dP_{k+1}}{dt} = \dfrac{\rho a^2}{F}\cdot\dfrac{Q_k - Q_{k+1}}{\Delta y} - \dfrac{\rho a^2}{F} q_k , \end{cases} \quad (2)$$

where the first equation represents the fluid transport in a (spatial) approximation element, and the second – conservation in approximation nodes, as elaborated in our previous publication [9]. These models form the bottom level of the ventilation models hierarchy (cf. Figure 4) and are used for the creation of the models of all upper hierarchical levels (airways/elements, sections, network), as shown in Figure 9. Connections between the models (or, precisely, their corresponding services) corresponded to the boundary conditions of the equations (2).

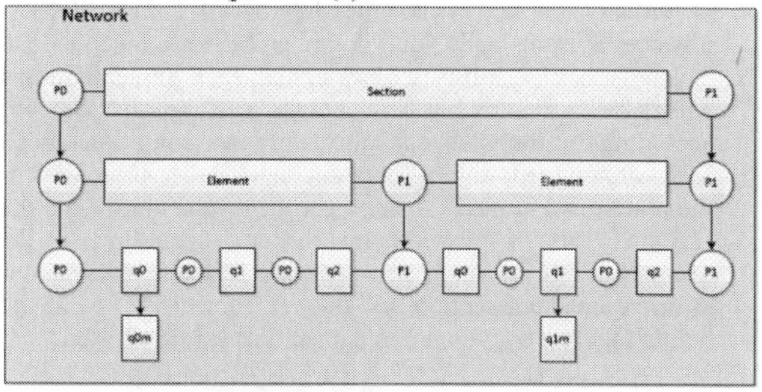


Figure 9. Hierarchical composition of modelling services.

The models were implemented with Open MPI as the underlying deployment and communication library. Three different scenarios were tested (A, B, C), as shown in Figure 10: an increase of fan pressure $P$ (scenario A), an adjustment of global regulator $r$ (scenario B), and a drop of fan pressure $P$ (scenario C). The results are shown for flow in the outbound element ($Q_{VS}$) and goaf ($q$), as well as marsh gas concentration $C$, which was defined as a ratio $qm/q$. All experiments (A, B, and C) showed results as expected to be by the physical experiments (e.g., performed by Svjatnyj in [5]). The service-based application workflow was steered by a dedicated supervising service – a manager, which performs the following functions:

- Initialization of microservices with the parameters that correspond to their respective physical object.
- Initiation of the iterative numerical solution process.
- Control of the solution readiness by means of polling the status of every individual "worker service".
- Instructing the services to store the results.

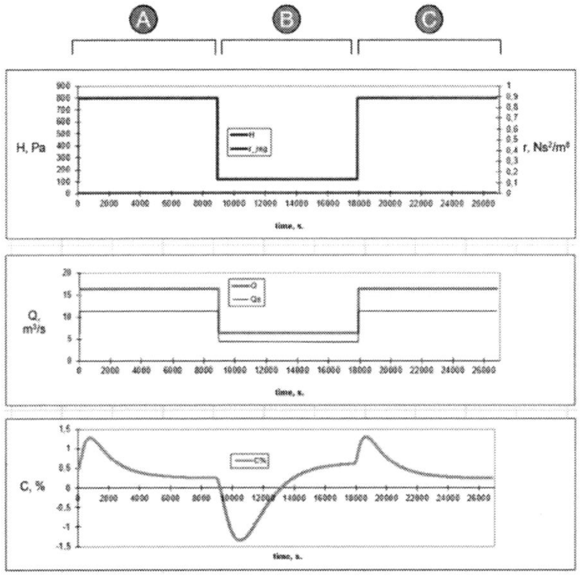


Figure 10. Modelling results.

The dynamic approach in which the services act as independent interactive components which are continuously running on the dedicated hardware and can be steered by a remote controller according to the specific application logic is particularly interesting for real-time control scenarios. In such scenarios, the services can incorporate the sensor data, make predictions for the future situation development, and instruct the control system about the probability of any potential risks appearance. On the other hand, the models can be optimized by adapting their parameters to best fit the actual mode of the controlled complex dynamic system.

## V. Conclusion

The simulation technology is facing the challenges of application for new real-time scenarios that require a high flexibility of modelling tools in terms of the broader usage of the available infrastructure (data acquisition, storage, and processing devices). The rapid development of sensor networks has made possible a number of new innovative scenarios, for which the monolithic design of the existing simulation tools and workflow solutions on their top might be a big obstacle. Service-oriented platforms offer a promising vision of the future development of simulation tools by offering benefits of on-demand distribution and parallelization, which might be well supported by the underlying management platforms. Microservices are the technology that can, if not fully replace the workflow-based scenarios, have the potential to support and bring them to the principally new level of usability. The effort that was done on implementation of the ventilation scenario has revealed a high potential of microservices architectures in geoscience and other domains of science and technology.

Further research will concentrate, among other things, on elaboration of functional composition strategies for development of complex, assembled services for hierarchically-organized systems.

## Acknowledgment

The work presented in this paper has become possible thanks to the support of the EU project ChEESE, which has received funding from the European Union's Horizon 2020 research and innovation program under the grant agreement N° 823844.

## References

[1] J. A. Kay, R. A. Entzminger, and D. C. Mazur, "Industrial Ethernet- overview and best practices", Conference Record of 2014 Annual Pulp and Paper Industry Technical Conference, Atlanta, GA, pp. 18-27, 2014.

[2] M. Broy, "Cyber-Physical Systems. Innovation through Software-Intensive Embedded Systems", Springer, 2010.

[3] P. Chang et al., "Cyberinfrastructure Requirements to Enhance Multi-messenger Astrophysics", In proc. Astro2020: Decadal Survey on Astronomy and Astrophysics, science white papers, no. 436; Bulletin of the American Astronomical Society, Vol. 51, Issue 3, id. 436, 2019.

[4] H. Andruleit et al., "BGR Energie study 2018 – data and development trends of German and global energy supplement", Bundesanstalt für Geowissenschaften und Rohstoffe (BGR), Hannover, 2019.

[5] V. Svjatnyj, "Simulation of airdynamic processes and development of control system for underground mine ventilation", PhD thesis 1985, (in Russian).

[6] C. Stewart, S. Aminossadati, and M. Kizil, "Use of live sensor data in transient simulations of mine ventilation models", Mining Report 153 (4/2017), pp. 356-363, 2017.

[7] A. Cheptsov, "From static to dynamic: a new methodology for development of simulation applications", chapter in Advances in Intelligent Systems: Reviews, Book Series, Vol.1, pp. 69-88, 2017.

[8] E. Clausen, "Mine ventilation in the 21st century – development towards adaptive ventilation systems", Mining Report Glückauf 153 (4/2017), pp. 326-332, 2017.

[9] A. Cheptsov, "The system organization and basic algorithms of the simulation and servicing center for the coal industry", in IEEE Proceeding International Conference Modern Problems of Radio Engineering,Telecommunications and Computer Science TCSET'2007, pp. 205-207, 2007.

[10] Wikipedia. Statistics of major catastrophes in coal mines, [Online]. Available from: https://de.wikipedia.org/wiki/ Liste_von_Ungl%C3%BCcken_im_Bergbau/ 2020.03.24

# HPC-Enabled Geoprocessing Services
# Cases: EUXDAT, EOPEN, and CYBELE European Frameworks

José Miguel Montañana
High Performance Computing Center
Stuttgart (HLRS) University of Stuttgart,
Nobelstraße 19, 70569
Stuttgart, Germany
Email: montanana@hlrs.de

Antonio Hervás
Inst. Matemática Multidisciplinar (IMM)
Universitat Politècnica de València
Camino de Vera s/n, 46020
Valencia, Spain
Email: ahervas@mat.upv.es

Dennis Hoppe
High Performance Computing Center
Stuttgart (HLRS) University of Stuttgart,
Nobelstraße 19, 70569
Stuttgart, Germany
Email: hoppe@hlrs.de

***Abstract*—There are big challenges with a great impact on the economy that can be addressed with geoprocessing such as the improvement in agricultural productivity, design of transport networks, prediction of natural disasters, or the study of climate change. This paper introduces recent developments in three European projects in High-Performance Computing (HPC)-Enabled geoprocessing Services applied to agricultural issues. The main goals of the European Union (EU) projects EUXDAT (*extreme data analytics in sustainable development*), CYBELE (*fostering precision agriculture and livestock farming through secure access to large-scale HPC-enabled virtual industrial experimentation environment empowering scalable big data analytics*), and EOPEN (*open interoperable platform for unified access and analysis of Earth observation data*) are, in general, to enable the use of large HPC systems, as well as big data management and user-friendly access and visualization of the results. In addition, these three projects focus on the development of software frameworks, develop Artificial Intelligence (AI) algorithms, and fuse Earth-Observation data, such as Copernicus data, and non-Earth-Observation data, such as weather, environmental and social media information. Finally, some initial results are shown.**

***Keywords–High-Performance Computing; Cloud Computing; Big Data; Agriculture; Land Monitoring; Machine Learning.***

## I. INTRODUCTION

Geoprocessing is a tool that allows addressing important and complex challenges. It is understood as the mathematical processing done by geographic Information Systems (GIS). During the last decades, the results of geoprocessing have greatly improved thanks to the exponential technological progress in computational power as well as exponential decreases in costs. The GIS systems consist essentially of three parts, as shown in Figure 1. The first part is that of data storage, the second part of computational processing, and the third part of visualization or access to results.

However, challenges such improve the efficiency of agricultural productivity require increasing by several orders of magnitude both the amount of data to be stored, as well as computational load. Therefore, the improvement in each of the aspects of geoprocessing presents itself as a new challenge.

The rest of this paper is organized as follows: Section II provides a summary of the contributions of this paper. Section III describes the implementation and Section IV the pilots and use cases. Sections V and VI describe the testing environment and comments on the experiments executed, respectively. Finally, conclusions are provided in Section VII.

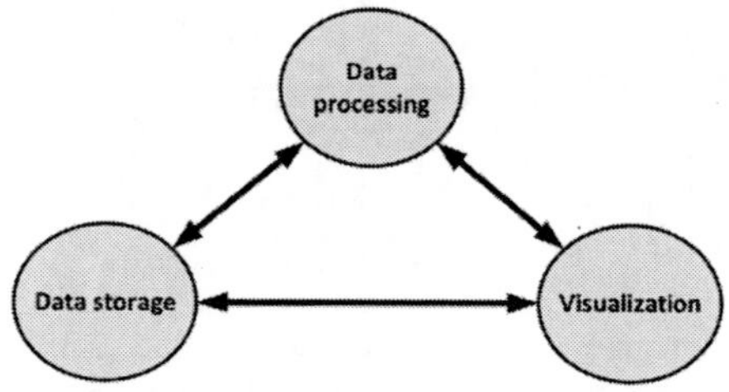


Figure 1. Fundamental components of geoprocessing systems.

## II. CONTRIBUTIONS OF THIS PAPER

The main contribution of this paper is to present an innovative platform for solving multiple technological challenges. Such challenges are the integration of data from different origins in different formats, the definition of interfaces for geoprocessing applications, the capability to run such applications on computing resources like HPC and in the cloud. Additionally, the platform faces the challenges of huge data transfers as well as enforcing secure access and permission control for the data and the computation results.

These challenges are targeted by the EU projects EUXDAT [1], EOPEN [2] and CYBELE [3]. In particular, these projects focus on developing solutions for the collection of big-data from different sources, data transfer into large-scale High-Performance-Computing centers and Cloud Computing for processing, as well as visualization services and access to the results.

Here, we provide a summary of the goals of these three projects:

### *A. EUXDAT*

EUXDAT proposes an e-Infrastructure for enabling Large Data Analytics-as-a-Service, which addresses the problems related to the current and future huge amount of heterogeneous data to be managed and processed within the agricultural domain. EUXDAT builds on existing mature components by providing an advanced frontend, where users will develop applications on top of an infrastructure based on HPC and Cloud. The frontend provides monitoring information, visualization, different distributed data analytic tools, enhanced data and processes catalogs. EUXDAT includes a large set of data connectors such as Unmanned Aerial Vehicles (UAVs), Copernicus, and field sensors for scalable analytics. Figure 2 shows the type of field sensors deployed for the EUXDAT project in farming areas. These weather stations [4] allow measuring a wide range of measurements on remote areas, like rain gauge, air temperature, air humidity, global radiation, wind speed, soil temperature, and leaf wetness.

Figure 2. Field sensors deployed in the farming areas.

As for the brokering infrastructure, EUXDAT aims at optimizing data and resource usage. In addition to a mechanism for supporting data management linked to data quality evaluation, EUXDAT proposes a way to orchestrate the execution of tasks, identifying whether the best target is HPC or Cloud. It uses monitoring and profiling information for making decisions based on trade-offs related to cost, data constraints, efficiency, and resource availability. During the project, EUXDAT is in contact with scientific communities, in order to identify new trends and datasets, for guiding the evolution of the e-Infrastructure. The result of the project will be an integrated e-Infrastructure, which encourages end-users to create new applications for sustainable development.

EUXDAT demonstrates real agriculture scenarios, land monitoring and energy efficiency for sustainable development, as a way to support planning policies.

### B. CYBELE

CYBELE is a European research project combining Agriculture, HPC, and Big Data. It involves 31 research institutes and enterprises across EU countries. It stands for: Fostering Precision Agriculture and Livestock Farming through Secure Access to Large-Scale HPC-Enabled Virtual Industrial Experimentation Environment Empowering Scalable Big Data Analytics.

CYBELE generates innovation and creates value in the domain of agri-food, and its verticals in the sub-domains of Precision Agriculture (PA) and Precision Livestock Farming (PLF) specifically, as demonstrated by the real-life industrial cases to be supported, empowering capacity building within the industrial and research community. The project aspires at demonstrating how the convergence of HPC, Big Data, Cloud Computing, and the Internet of Things (IoT) can revolutionize farming, reduce scarcity and increase food supply, bringing social, economic, and environmental benefits. It develops large scale HPC-enabled testbeds and delivers a distributed big data management architecture and a data management strategy.

### C. EOPEN

The objective of EOPEN is to fuse Earth Observation (EO) data with multiple, heterogeneous and big data sources, to improve the monitoring capabilities of the future EO downstream sector. The Earth Observation data consists of the Copernicus and Sentinel data, while the non-EO data is weather, environmental and social media information.

The fusion is done at the semantic level, to provide reasoning mechanisms and interoperable solutions, through the semantic linking of information. Processing of large streams of data is based on open-source and scalable algorithms in change detection, event detection, data clustering, which are built on High-Performance Computing infrastructures.

Alongside this enhanced data fusion, a new, innovative architecture, overarching Joint Decision & Information Governance, is combined with the technical solution to assist with decision making and visual analytics. EOPEN is demonstrated through real use case scenarios in flood risk monitoring, food security, and climate change monitoring.

## III. IMPLEMENTATION

The main goal is to develop a sustainable approach that facilitates access to data and geoprocessing applications and, at the same time, using the state-of-the-art on big-data management, as well as computation resources from Cloud platforms to HPC centers.

The target of the implementation is to provide an open-source system that can be used by commercial products, as well as by other projects to run after being finalized. The reason for including support for accounting and billing is to facilitate the code actually being used in the future since it is necessary to consider the costs of using large computer systems, as well as the cost of data acquisition from proprietary sources.

Therefore, each of the components has a clearly defined User-Interface. Figure 3 shows the main components of the infrastructure platform. The first one is the User-Interface (UI) Application Programming Interface (API). This supports the development of applications such as mobile devices or web-interfaces, without knowing the complexity of the other components.

The portal provides users with a list of available applications and the data catalog available for them. The users do not need to consider the complexity or format of the data, neither the different data sources, because it is encapsulated by the platform and the applications internally.

The data catalog collects data from different data sources. Some of these sources are free while others have an economic cost. Similarly, the catalog of applications can use free applications such as those developed in this project, or commercial applications. This is done to raise interest in using the platform by third parties that wish to commercialize applications or data.

Thus, once users select the task to perform, such as the prediction of temperature for a particular field on a particular date, they just wait for the result. Notice that the computation time is reduced by multiple orders of magnitude when using a large-scale HPC system.

The user request is submitted to the orchestrator, which is responsible for the transfer and execution of the applications on

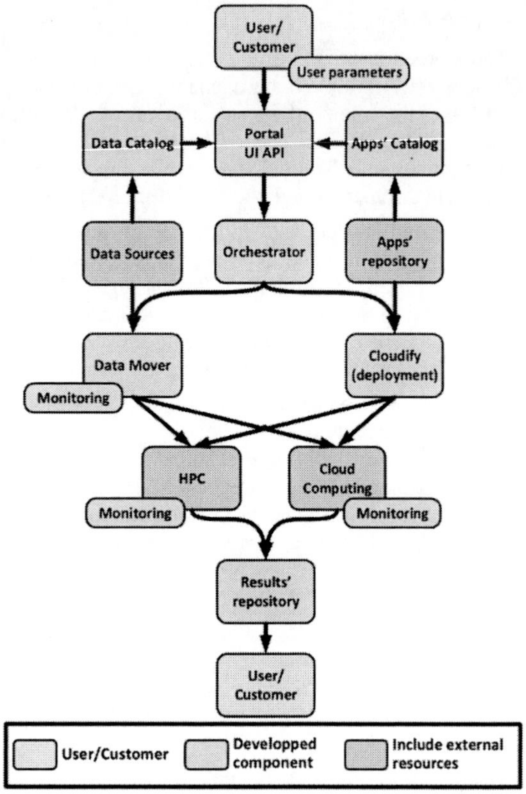


Figure 3. Infrastructure platform.

the computing resources. Basically, the orchestrator, based on the user request, selects the appropriate computation resource such as HPC or Cloud. Also, it queues a blueprint file [5] with the specifications of the user parameters, the input and output files, as well as the binary files to be transferred into the computation resources. A fragment of a blueprint is shown in Figure 4. Thus, the blueprint file allows the orchestrator which receives workload requests to delegate the required staging of input and output data to the Data-Mover component.

```
...
node_templates:
job:
        type: croupier.nodes.job
        properties:
        job_options:
                type: "SRUN"
                command: "olu coordinates.txt"
                nodes: 100
                max_time: "04:00:00"
        deployment:
                bootstrap: "bootstrap.sh"
                revert: "revert.sh"
                inputs:
                - "first_job"
                - {get_input: part_1}
...
```

Figure 4. Example of a fragment of a blueprint file.

The transfer of files is controlled by the Rucio server [6]. Rucio is the state of the art on large-scale data management. It is open-source and developed by ATLAS [7] for managing big-data at the European Organization for Nuclear Research (CERN); it is currently used to move more than 1 petabyte per day, and more than one million files per day [8][9].

Rucio allows defining three levels of architecture of file access in the DataMover. The lowest level is the storage of data in physical storage systems, These storage systems are referred to as Rucio-Storage-Elements (RSEs). The intermediate level corresponds to the logical access to the files. The physical location of the file is obtained from a database based on the logical identifier of the file, which consists of a text label. Thus, it is not necessary to provide the physical location of the requested files. At the highest level, datasets or sets of files are defined. Note that physical files can be included logically in different datasets without the need to be physically replicated. This makes it possible to avoid transmitting the same file multiple times to the same destination, for example, if an input file was copied to a certain computer system, it would not need to be transmitted again for any other application that needs it. In particular, we set up an RSE on the computation side. Rucio allows uploading data there and, at the same time enforces secured access and permission control for those files.

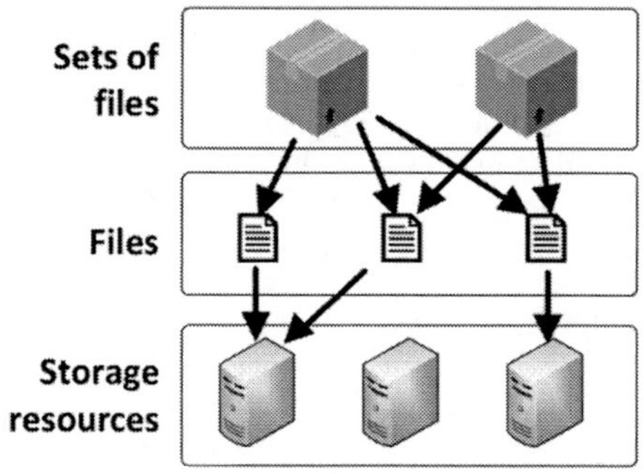


Figure 5. Three layers for data access.

In order to improve future application executions, the utilization metrics of the different resources are registered into the monitoring Prometheus server [10]. This will help with the decision on where to allocate the next requests depending on the user constraints, such as reducing computation time or reducing computation cost. Once the computation is completed, the results are moved into an accessible repository by the end-user, and the user is notified.

## IV. USE CASES

The three projects presented above are focused on the development and test solutions for the agriculture field. Agriculture is a key aspect of economic and political stability. Because of its importance, governments are funding the development of solutions for those challenges based on data access systems, geoprocessing, and support for decision making.

The different uses cases will demonstrate the capacity of the HPC solutions proposed in the projects. They will be eventually open for end-users communities in the last phase of the projects, but currently, only consortium partners have access to the pilots' implementation.

The use cases cover a wide range of scenarios from detecting weather conditions, humidity or crop diseases up to Precision Agriculture, Livestock Farming, and exploration. Here, we provide a brief description of some of them.

**Pilot for Open-Land Monitoring and Sustainable Management Implementation**: It targets on developing a deep learning algorithm which correlates input spectral data with ground truth, to be used for prediction of soil and crop status. To achieve it, multi-rotor UAV systems with a hyperspectral camera combined with earth-observation and meteorological data will be used for classification of crop status.

**Pilot for Energy Efficiency Implementation**: It focuses on developing analytics algorithms in order to obtain models of processes cost and profits to support energy-efficiency in agriculture.

**Pilot for 3D Farming Implementation**: It focuses on analytics models, mainly, on spatial analysis, for locating the highest productivity zones. It will provide 3D visualization for the obtained results, which especially help to understand the conditions of water, soil particles, and nutrients.

**Organic Soya yield and protein-content prediction**: There is an interest in the prediction of the soybean cultivation, due to the EU is strongly dependent on other continents for plant-based proteins. For that reason, this use case develops methods for predicting yield and protein-content maps based on crowdsourced data, satellite imagery and additional information, when available, such as electromagnetic soil scans, and other sensory data.

**Climate-Smart Predictive Models for Viticulture**: It targets the development of complex, highly-nonlinear models for vine and grape growth, which rely on a large number of variables that have been shown to affect the quality and quantity of the produced yields. The range of data includes earth observations, soil/elevation maps, genomics data, chemical analysis data, environmental and climatic data.

**Climate services for organic fruit production**: The goal is to help with the prevention of damage effects due to frost and hail. The solution under development focuses on providing risk probability mapping calculated based on models obtained by machine learning techniques. To do that, a wide range of data sources is used including but not limited to climate instability indices, digital terrain models, in-situ environmental and climatic data, and satellite images.

**Optimizing computations for crop yield forecasting**: Crop yield monitoring can be used as a tool for agricultural monitoring (e.g. early warning & anomaly detection), index-based insurance (index estimates) and farmer advisory services. Its goal is to compute a productivity estimation based on cropping systems model and a combination of different datasets, such as ingest crop, soil, historic weather data, weather forecasts data. However, it becomes a challenge to do that computation as the amount of available data keeps increasing as well as it is resolution.

## V. EXPERIMENTS

In order to test the platform, we have been testing the deployment of the EUXDAT software platform in Hazelhen supercomputer at HLRS. Table I shows the details of the current supercomputer [11], and the new one to be installed on Q1 2020 [12].

The simultaneous use of large systems by a large number of users requires that each user execution request has to specify

TABLE I. CHARACTERISTICS OF THE HLRS SUPERCOMPUTERS.

| Name | Cray XC40(HazelHen) | HPE Apollo 9000(Hawk) |
|---|---|---|
| Number of cores | 185,088 | 720,896 |
| Storage capabilities | 10 PB* | 25 PB* |
| Interconnection network | Ariel | InfiniBand HDR (200Gbit/s) |
| Power consumption | 3200 KW | Initially 3200 KW, but planned to be increased |

*: 1PB = 1024 TB = 1,048,576 GB = 1,073,741,824 MB

the number of computing nodes and software to be used. The requested executions will keep waiting in a queue until there will be free resources to fit the particular requirements of each one. Notice that the waiting time can be from a few minutes to a few days depending on the load on the supercomputer. Obviously, the cost to bill the user will be based only on the effective computation time. The cost never includes the queue waiting time. Therefore, the required computation time is an important aspect of using geoprocessing applications in real cases.

The platform proposed in this paper uploads the required data in advance and not during the computation time. Because uploading data in advance can save significant computation time for geoprocessing applications. And therefore, it saves a significant cost. Notice that each of the use cases is composed of a series of geoprocessing applications. Thus, the computation time and data storage requirements of each use case will be the accumulation of the requirements of those applications. We can not provide the final requirements of the geoprocessing applications in our projects, because their implementation and parallelization are still not finalized. For that reason, we provide here only the preliminary requirements of the two applications which are commonly shared in almost all of the listed use cases. Table II shows the current data size to transfer and the required computation load of the applications for calculation of the land morphometry characteristics and weather predictions.

TABLE II. REQUIREMENTS OF TWO DIFFERENT APPLICATIONS.

| Applications | Agroclimatic zones Frost date calculation | Morphometry characteristics calculation |
|---|---|---|
| Storage requirements | 316 MB (ERA5-land Czech Rep) | 25 GB (Austria Area) 1 TB (Full Europe) |
| Computation time in core-hours | 70 (Czech Republic) | 3000 (Full Europe) |

The computation time is also an important aspect to take into account when there is a need to have the result at a certain time. For instance, a farmer needs to know if the next morning's temperature is below 27 degrees, because "at full bloom, the blossoms are usually killed by temperatures around 27 degrees" [13]. We consider that in these cases is preferred to use a computer center. Because in our experience, computing on the Cloud takes an immense amount of time when compared with the computation at the HLRS supercomputer.

The benefit of time-efficient computing on a supercomputer requires that the application be prepared to run in parallel. However, there was not needed a big effort to prepare the first parallelization of geoprocessing applications like the morphometry characteristics calculation. Because in this particular case, the load was easily distributed among computing nodes just by dividing the computation load by geographical areas to be processed. Currently, considering that the applications are implemented with python, the developers' team is using Message Passing Interface (MPI) for Python [14], since it seems the most efficient way to obtain the best performance on these systems.

## VI. Analysis of the Experience

The proposed platform currently satisfies all use case requirements, and there were not deficiencies detected. The proposal simplifies the deployment and execution of geoprocessing tasks. It helps to do a more efficient deployment of data and computation, both in terms of time. The experience shows that the proposal seems to be the best cost-effectiveness for geoprocessing, especially for big projects, in particular for governmental large scale studies. In addition, it seems that the proposal is a cost-effective solution for companies interested in selling results of geoprocessing to small customers that do not have access to the data or the software to do the computation by themselves.

## VII. Conclusions

In this paper, the infrastructure for HPC and Cloud computing of geoprocessing services has been described. The infrastructure is running and the use cases are under the last development stages in the last year of the projects EUXDAT and EOPEN, while CYBELE will keep running until the end of 2021.

The solutions being developed will greatly support improving farming performance and competitiveness, not only providing access to the tools, but also because the tools will run on most time-efficient computation resources. They will simplify the access for non-technical users, such as farmers who may access the services through their mobile phones. The developed platforms are expected to keep running after the end of the project. The partners in the projects are interested to use them for selling their products, such as datasets and weather forecasting services. For that reason, the EUXDAT consortium is looking for attracting service providers that sell the final products and services directly to the farmers. The consortium can potentially take the roles of software and cloud platform provider in order to support the ASPs.

Another aspect is that the developed platform for agriculture geoprocessing is also suitable for other purposes than agriculture, such as providing optimum paths through transportation networks, or predicting disasters like wildfire, flooding, or effects of a storm. Potential users can also include local authorities interested in Urban and Regional Planning and water management, or insurance companies interested in risk prevention or disaster resilience.

## Acknowledgment

This work has been done within the projects *European e-infrastructure for extreme data analytics in sustainable development* (EUXDAT), *fostering precision agriculture and livestock farming through secure access to large-scale HPC-enabled virtual industrial experimentation environment empowering scalable big data analytic* (CYBELE), and *Open interoperable platform for unified access and analysis of Earth observation data* (EOPEN). See the projects' web pages [1][2][3] for further information. The research leading to these results has received funding from the European Unions Horizon 2020 Research and Innovation Programme, grant agreements n. [777549, 825355, 776019], respectively.

We wish to especially thank the partners who have collaborated with this paper providing their applications, as well as the estimation of their requirements. Those are Dimitrij Kozuch (P4ALL), Pavel Hájek (WRLS), and Dr. Karl G. Gutbrod (CEO at Meteoblue AG). We would also like to thank the work and collaboration of the rest of more than 31 research institutes and enterprises across EU countries partners in these projects. The list is too long to mention everyone.

## References

[1] F. J. Nieto et al. (EUXDAT consortium), "EUXDAT *European e-Infrastructure for Extreme Data Analytics in Sustainable Development*," [online]: https://www.euxdat.eu/ [retrieved: Mar-2020].

[2] G. Vingione et al. (EOPEN consortium), "EOPEN *Open interoperable platform for unified access and analysis of earth observation data*," [online]: https://eopen-project.eu/ [retrieved: Mar-2020].

[3] S. Davy et al. (CYBELE consortium), "CYBELE *Fostering Precision Agriculture And Livestock Farming Through Secure Access To Large-Scale Hpc-Enabled Virtual Industrial Experimentation Environment Empowering Scalable Big Data Analytic*," [online]: https://www.cybele-project.eu/ [retrieved: Mar-2020].

[4] Pessl Instruments GmbH, "Stations and dataloggers," 2020, [online]: http://metos.at/micrometos-clima/ [retrieved: Mar-2020].

[5] J. Carnero, "Example of blue-print, available on the github project," 2019, [online]: https://github.com/hlrs-121991-germany/croupier/blob/master/croupier_plugin/tests/blueprints/blueprint_four.yaml [retrieved: Mar-2020].

[6] "Rucio: scientific data management," 2020, [online]: https://rucio.cern.ch/ [retrieved: Mar-2020].

[7] "Atlas experiment," [online]: https://atlas.cern/discover [retrieved: Mar-2020].

[8] C. Serfon et al., "Rucio, the next-generation Data Management system in ATLAS," Nuclear and Particle Physics Proceedings, vol. 273–275, 2019, p. 969–975, [retrieved: Mar-2020].

[9] R. Gardner, B. Riedel, and M. Lassnig, "Rucio concepts and principles," Dec. 2017, Presentation available at URL: https://indico.fnal.gov/event/15861/session/0/contribution/2/material/slides/0.pdf [retrieved: Mar-2020].

[10] "Prometheus sorftware: free software application for event monitoring and real-time alerting," 2020, [online]: https://prometheus.io/ [retrieved: Mar-2020].

[11] University of Stuttgart, HLRS, "Technical Description of the Hazelhen Supercomputer," 2020, [online]: https://www.hlrs.de/systems/cray-xc40-hazel-hen/ [retrieved: Mar-2020].

[12] ——, "Technical Description of the Hawk Supercomputer," 2020, [online]: https://www.hlrs.de/systems/hpe-apollo-9000-hawk/ [retrieved: Mar-2020].

[13] J. Muhollem, "Warm winter has put state's apple crop at risk, expert warns," 2017, Pennsylvania State University, [online]: https://phys.org/news/2017-03-winter-state-apple-crop-expert.html [retrieved: Mar-2020].

[14] L. Dalcin, "MPI for Python," 2019, [online]: https://mpi4py.readthedocs.io/en/stable/intro.html [retrieved: Mar-2020].

# Automatic Publication of Open Data from OGC Services: the Use Case of TRAFAIR Project

Javier Nogueras-Iso*, Héctor Ochoa-Ortiz*, Manuel Ángel Jañez*, José R. R. Viqueira†, Laura Po‡ and Raquel Trillo-Lado*
*Universidad de Zaragoza, Spain
Email: {jnog,719509,731321,raqueltl}@unizar.es
†Universidade de Santiago de Compostela, Spain
Email: jrr.viqueira@usc.es
‡Università degli Studi di Modena e Reggio Emilia, Italy
Email: laura.po@unimore.it

*Abstract*—**This work proposes a workflow for the publication of Open Spatial Data. The main contribution of this work is the automatic generation of metadata extracted from Open Geospatial Consortium (OGC) spatial services providing access to feature types and coverages. Besides, this work adopts a geospatial extension of the Data Catalog Vocabulary metadata application profile for data portals in Europe for the description of datasets. This extension, called GeoDCAT-AP, has been adopted because it allows for an appropriate crosswalk between the annotation requirements in the spatial domain and the metadata models accepted in general Open Data portals. The feasibility of the proposed workflow has been tested within the framework of the TRAFAIR project to publish monitoring and forecasting air quality data.**

***Keywords–Environmental data; Open Data; GeoDCAT-AP; metadata; geospatial services; OGC.***

## I. Introduction

TRAFAIR (Understanding traffic flows to improve air quality) is a European project co-financed by the Connecting Europe Facility of the European Union (Project Nr. 2017-EU-IA-0167), whose main objectives are the monitoring of air quality in urban areas, and the development of forecasting air quality services based on meteorological predictions and urban traffic flows [1], [2]. The project also aims to publish monitoring and forecasting air quality data as Open Data and to develop client applications to make both citizens and public administrations aware of the air quality and the responsible use of private transport.

To facilitate the visualization and download of monitoring and forecasting data, project partners have chosen the use of GeoServer software, which facilitates the setting up of servers accessible through the standardized service interfaces compliant with Open Geospatial Consortium (OGC) specifications: Web Mapping Services (WMS) for visualization, Web Feature Services (WFS) for the download of feature data, and Web Coverage Services (WCS) for the download of coverage data. Figure 1 shows a layered architecture with the data and service components managed in the TRAFAIR project.

However, the simple publication of OGC services cannot be considered as Open Data publication. To make data really accessible as Open Data, we need to register datasets in official Open Data portals. Furthermore, the publication of datasets in the European Data Portal (EDP) [3] is a requirement of the project. To register as Open Data the TRAFAIR outcomes, the first step has been to select an appropriate metadata profile compliant with the metadata models accepted in the Open Data context. Taking into account the spatial character of data managed in TRAFAIR, we have adopted the GeoDCAT-AP metadata profile [4]. GeoDCAT-AP is a metadata profile that extends DCAT-AP, a metadata profile designed by the European Commission to describe public sector data. GeoDCAT-AP metadata properties have been designed to assure compliance with the metadata requirements of the European INSPIRE directive for establishing a spatial information infrastructure in Europe [5].

On the other hand, to minimize the effort of creating metadata and the registration of these data in Open Data portals, we decided to automate this process employing a software that retrieves the capabilities of OGC services and converts this information into metadata records that are later ingested in a CKAN-like Open Data server. CKAN [6] is the most widely used Open source platform to support Open Data portals, which includes the necessary plug-ins to exchange metadata in RDF format (the serialization format used for GeoDCAT-AP).

The objective of this work is to describe the workflow that we have proposed for the publication of Open Spatial Data integrating the automatic generation of GeoDCAT-AP metadata. The remainder of this paper is structured as follows. Section II introduces background information on the GeoDCAT-AP metadata model. Section III describes our proposed workflow for the publication of Open Spatial Data. Section IV describes the feasibility of the application of the proposed workflow in the cities of Modena, Santiago de Compostela, and Zaragoza. Section V reviews related works in the literature. Last, this paper ends with some conclusions and an outline of future work.

## II. GeoDCAT-AP: A Metadata Profile for Open Spatial Data

ISO 19115 is the international standard for geographic metadata proposed by the International Organisation for Standardization (ISO) [7], which has been widely adopted during the last decade in the geographic information community in both public and private sectors.

However, in the Open Data domain, more general and simple metadata schemas are needed to facilitate the publication of datasets from different disciplines in the same metadata

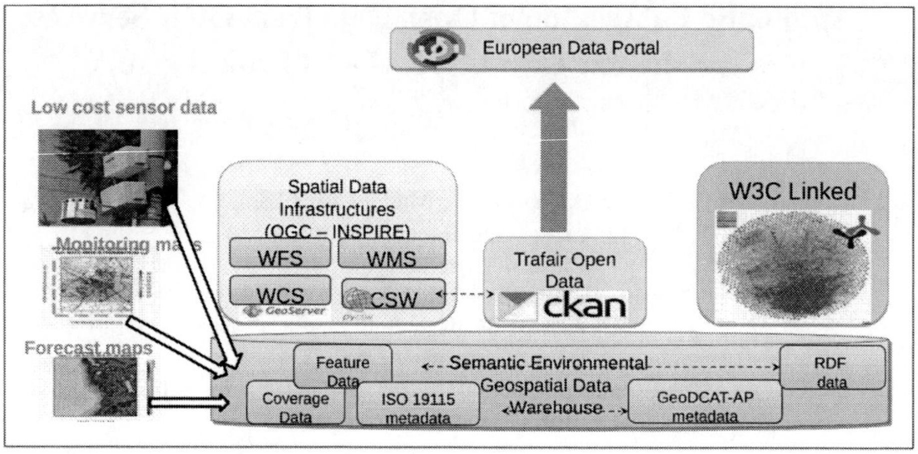


Figure 1. Architecture of data and services components in TRAFAIR project

repository. DCAT is the acronym for W3C's Data Catalogue vocabulary) [8] and can be considered as a basic and general core of metadata properties shared by the different metadata schemas used in various Open Data initiatives. In the case of Europe, the European Union proposed in 2013 DCAT-AP [9], a specification based on DCAT for describing public sector datasets in Europe. Compared to DCAT, DCAT-AP provides stricter definitions of catalogs, datasets, distributions, and other objects.

As mentioned in the introduction, within the context of this project, we have selected GeoDCAT-AP v1.01 [4]. This extension of DCAT-AP [9] was designed for the description of spatial data and its metadata properties have an exact mapping with the main elements of ISO 19115 metadata. This mapping assures the transformation of GeoDCAT-AP records into equivalent ISO 19115 metadata records compliant with INSPIRE requirements [10].

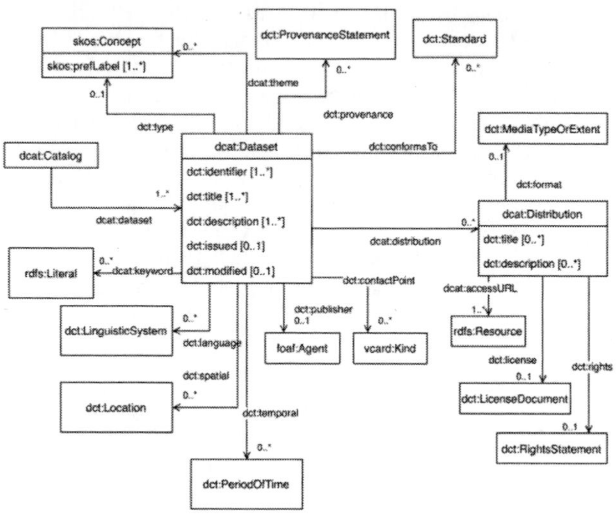


Figure 2. Entities and properties used from GeoDCAT-AP

The description of datasets according to GeoDCAT-AP is mainly focused on providing information about three main entities: a *Catalog* that is published through an Open Data portal containing *Datasets* and the associated *Distribution* forms of each dataset. Besides, GeoDCAT-AP makes a distinction between core and extended properties. The core set is the selection of DCAT-AP metadata properties that have a direct binding with ISO 19115 and INSPIRE metadata. The extended set is a superset of the core set, including additional metadata properties to provide a complete binding with ISO 19115 and INSPIRE metadata. In some cases, these additional properties belong to other metadata vocabularies. In other cases, although the properties belong to DCAT-AP, they are classified as extended because they only provide a partial binding with ISO 19115 and INSPIRE.

Figure 2 shows a UML diagram with the properties from GeoDCAT-AP that are needed for describing datasets and distributions in TRAFAIR. Most of these properties belong to the core set of GeoDCAT-AP. The only exceptions are the *dct:type* property of *Datasets* and the *dct:description* property of *Distributions*. *dct:type* is employed to indicate whether the described resource is a dataset or a dataset series. *dct:description* allows the description of the spatial resolution of associated distributions. Although GeoDCAT-AP proposes *rdfs:comment* as a provisional property to fill this resolution information, there is no direct mapping of this property to CKAN fields and we have considered *dct:description* as a valid alternative.

## III. Proposed Workflow for the Publication of Open Spatial Data

Figure 3 shows an activity diagram with the main five steps of the workflow that we have proposed for the publication of Open Spatial Data. For steps 1, 3, and 4, we have developed software to automate as much as possible the automatic generation and release of metadata. Steps 2 and 5 are accomplished thanks to the use of existing software packages.

The first step is the ingestion of layers in a spatial data warehouse. GeoServer is the software package selected for managing the publication of spatial data layers, either discrete feature data or coverage data. In the context of this project, we have developed specific software in Java and R languages to ingest feature types (supported in a spatial database) and

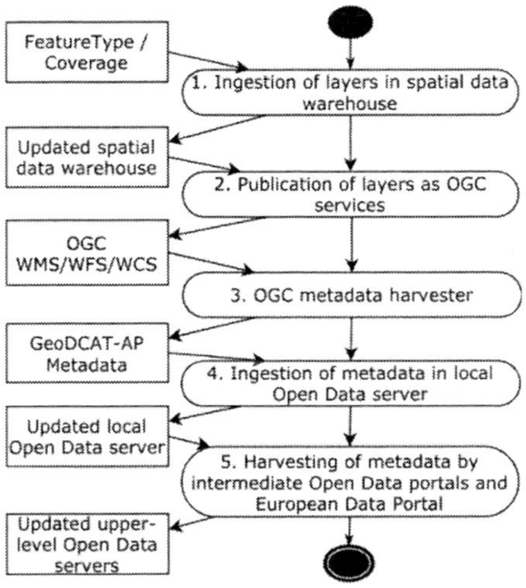


Figure 3. Workflow for publication of Open Spatial Data

coverages in GeoServer through its REST API [11], [12]. Concerning metadata generation, this software takes care of sending to the REST API the appropriate values for the tags enumerated in the *GeoServer* column of Table I.

The second step is the publication of layers as OGC services. This step is directly achieved thanks to the GeoServer software, which provides access to layers through different OGC services. Feature types, such as observations retrieved from traffic and air quality sensors, may be downloaded through a WFS service. In the case of coverages for air quality monitoring (interpolations of geo-referenced sensor observations) or coverages for predicting quality (the result of applying a Lagrangian model for the dispersion of pollutants called GRAL [13]), a WCS service is used to retrieve these raster data. Beyond WFS and WCS, some layers are also available to perform server-side map rendering using a WMS service.

The third step is the harvesting of metadata from OGC services through its *GetCapabilities* operation. To implement this step, we have developed a Python program that takes profit of OWSLib [14], a Python package for client programming with OGC web service interface standards and their related content models. This software interacts with the OGC interface, instead of the GeoServer REST API, because we wanted to make this software scalable enough to integrate in the future other layers managed by software packages different from GeoServer. The algorithm behind this software generates a *Dataset* instance for every layer published in WFS or WCS services. In addition, each *Dataset* has at least one associated *Distribution* instance in the form of a link to a WFS or WCS service. In some cases, if the layer is also rendered through a WMS, a second distribution linking to the WMS service is generated. The *OWSLib* column in Table I shows the fields retrieved with OWSLib package to generate the corresponding GeoDCAT-AP property.

The fourth step is the ingestion of metadata in the Open Data server of the institution in charge of publishing the air quality data of the local area. As a continuation of the software in the previous step, our Python program transforms the information retrieved in the previous step into a dictionary with the required items to construct a dataset and its associated resources, which are immediately inserted in the CKAN-based local Open Data server through its REST API [15]. The *CKAN* column in Table I indicates the tags that are used in this dictionary data structure to generate later RDF metadata based on GeoDCAT-AP. The mapping between CKAN fields and RDF properties is the one proposed in the *ckanext-dcat* plugin of CKAN [16].

The final step is the harvesting of metadata in the local servers by regional and national Open Data portals until the EDP finally harvests metadata. This step is beyond the scope of the TRAFAIR project. However, we assume that upper-level portals are based on CKAN technology (or have a similar mechanism for the harvesting of subscribed lower level catalogs). On the one hand, the *ckanext-dcat* plugin of CKAN allows the publication of datasets metadata as RDF in compliance with DCAT-AP vocabularies. On the other hand, the *ckanext-harvest* plugin of CKAN allows us to harvest the contents of different types of catalog sources.

Last, it must be noted that the steps of the workflow can be either executed automatically in a row, or they can be interleaved with manual supervision to revise the information associated to layers in GeoServer (before applying steps 3 and 4) or the metadata in local CKAN servers (before harvesting takes place in step 5). Besides, the workflow can be applied incrementally to take into account new layers created in GeoServer, or to update CKAN metadata if the configuration of layers in GeoServer has changed.

## IV. Deployment of Open Data in the Cities of Modena, Santiago de Compostela and Zaragoza

Figure 4 shows the deployment of specific Open Data portals in the cities of Modena (Italy), Santiago de Compostela (Spain) and Zaragoza (Spain). The figure also shows how the local GeoServers are queried with the software described in steps 3 and 4 of the proposed workflow to feed the contents of the local Open Data portals. In addition, the figure shows the Open Data portals at regional, national, and European level that harvest the contents of the local Open Data portals.

In the case of Modena, the Open Data contents managed by the municipal government of Modena (*Comune di Modena*) are directly ingested in the CKAN-based Open Data server provided by the regional government of Emilia-Romagna [17]. The contents of this portal are harvested by the Italian Government Open Data portal (*dati.gov.it*).

In the case of Santiago de Compostela, the Open Data portal is managed by the municipal government of Santiago de Compostela (*Concello de Santiago*) [18]. Later, the contents of this portal are harvested by the Spanish Government Open Data portal (*datos.gob.es*).

The case of Zaragoza is more complicated. There is an Open Data portal based on CKAN maintained by the researchers of the University of Zaragoza involved in the TRAFAIR project. However, the Open Data contents of the University are published through a different kind of repository (called Zaguan) based on MARC metadata and accessible through the OAI-PMH protocol. In this case, we had to develop a specific program to upload the GeoDCAT-AP metadata periodically in MARC format at Zaguan portal (see figure

TABLE I. CORRESPONDENCE BETWEEN GEOSERVER TAGS (CONTAINED IN THE BODY OF A POST REQUEST TO CREATE A FEATURETYPE/COVERAGE), LAYER FIELDS RETRIEVED WITH OWSLIB FROM A GETCAPABILITIES RESPONSE, CKAN TAGS (CONTAINED IN THE BODY OF A POST REQUEST TO CREATE A DATASET), AND GEODCAT-AP PROPERTIES

| GeoServer | OWSLib | CKAN | GeoDCAT-AP |
|---|---|---|---|
| featureType/name, coverage/name | *layerName* | extra:identifier | Dataset/dct:identifier |
| featureType/title, coverage/title | contents[*layerName*].title | title | Dataset/dct:title |
| featureType/description, coverage/description *(software in step 1 introduces predefined descriptions according to name patterns)* | contents[*layerName*].abstract | notes | Dataset/dct:description |
| | *("series" for OGC services with temporal dimension, or "dataset" without temporal dimension)* | extra:dcat_type | Dataset/dct:type |
| | *(default language proposed in step 3)* | extra:language | Dataset/dct:language |
| | *(default INSPIRE data themes and ISO 19115 topic categories proposed in step 3)* | extra:theme | Dataset/dcat:theme |
| *(some default keywords are automatically introduced by GeoServer)* | contents[*layerName*].keywords | tags | Dataset/dcat:keyword |
| *(computed automatically by GeoServer)* | contents[*layerName*].boundingBoxWGS84 | extra:spatial | Dataset/dct:spatial |
| *(start date and end date are automatically updated by GeoServer)* | contents[*layerName*].timepositions | extra:temporal_start + extra:temporal_end | Dataset/dct:temporal |
| | | extra:issued *(automatically inserted with first ingestion in CKAN)* | Dataset/dct:issued |
| | | extra:modified *(automatically updated with every update of a dataset in CKAN)* | Dataset/dct:modified |
| | *(default provenance proposed in step 3)* | extra:provenance | Dataset/dct:provenance |
| | *(default INSPIRE conformance and coordinate reference system proposed in step 3)* | extra:conforms_to | Dataset/dct:conformsTo |
| *(contact information is directly introduced by administrators at GeoServer configuration page)* | contents[*layerName*].provider.contact. organization | extra:publisher_name | Dataset/dct:publisher |
| | contents[*layerName*].provider.contact.name + contents[*layerName*].provider.contact.email | extra:contact_name + extra:contact_email | Dataset/dcat:contactPoint |
| *(OGC service URL generated automatically by GeoServer)* | *(OGC service URL)* | resource:url | Distribution/dcat:accessURL |
| featureType/name, coverage/name | *layerName* | resource:name | Distribution/dct:title |
| featureType/serviceConfiguration, coverage/serviceConfiguration | *("wfs", "wcs" or "wms" according to OGC service type)* | resource:format | Distribution/dct:format |
| | *(default licence proposed in step 3)* | resource:license | Distribution/dct:license |
| | *(default rights proposed in step 3)* | resource:rights | Distribution/dct:rights |
| | *(default resolution proposed for project datasets)* | resource:description | Distribution/dcat:description |

5) [19]. Then, these metadata records are harvested by the regional government of Aragon, and later by the Spanish Government Open Data portal.

All datasets provided by different TRAFAIR partners on local portals will also appear at EDP [3]. The EDP will act as a common collector of all data related to air quality and traffic that will be generated and published within the different cities and will also encourage the reuse of the TRAFAIR outcomes.

## V. RELATED WORK

There are several examples of works trying to crawl the contents of OGC services and automate the generation of metadata items that are later ingested in catalogs compliant with the OGC Catalog Services for the Web (CSW) specification. For instance, Nogueras-Iso et al. [20] proposed a mechanism to derive metadata from the capabilities information returned by OGC services (e.g., WMS, WFS or WCS) and create entries in a catalog of geographic information services. A related solution is the CSW - ISO 19115 community module of GeoServer software [21]. This GeoServer extension allows the browsing of GeoServer layers through a CSW API, but no details are provided about the OGC services providing access to the layers. This extension is also comparable to the harvesting possibility offered by Geonetwork (a software for deploying geographic metadata catalogs) to use the GetCapabilities response of an OGC service (e.g., WMS, WFS or WCS) to generate ISO 19115 metadata for the resources delivered by the service [22]. Another example studying in more detail the layers advertised in a *GetCapabilities* response is the one proposed by Florczyk et al. [23]. This work describes a method for the automatic detection of the orthoimage layers discovered in the *GetCapabilities* responses of Web Map Services, which was used to feed the contents of a virtual catalog of orthoimages.

Concerning the synchronized publication of data and metadata, there are also examples of works trying to define workflows for the joint publication of datasets and metadata. For instance, Gil-Altaba et al. [24] proposed a service framework that used the GeoServer REST API to create a new data store accessible through OGC services, and immediately ingest the associated metadata to this datastore into a CSW catalog supported with Geonetwork software.

The previous works are focused on generating ISO 19115 - compliant metadata. However, for publication of geographic information as Open Data, solutions generating DCAT-based metadata are required. Perego et al. [25] describe uses cases for profile-based content negotiation and publishing metadata on the web where a GeoDCAT-AP API has been developed to transform original content according to ISO 19115 metadata standard into DCAT-AP metadata in various formats. A similar approach is provided through the *ckanext-spatial* plugin of CKAN [26]. This plugin allows harvesting the ISO 19115 contents of CSW catalogs. Nevertheless, none of these two approaches derive metadata automatically from services.

The workflow for the publication of open data proposed in this work contributes to the state of the art as it provides

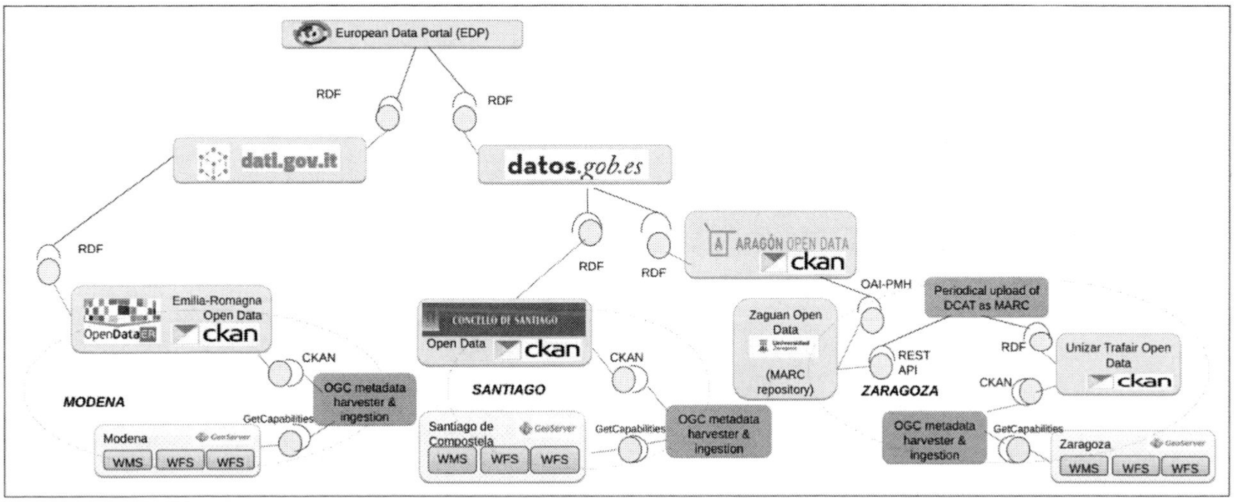


Figure 4. Deployment of Open Data servers in the cities of Modena, Santiago and Zaragoza

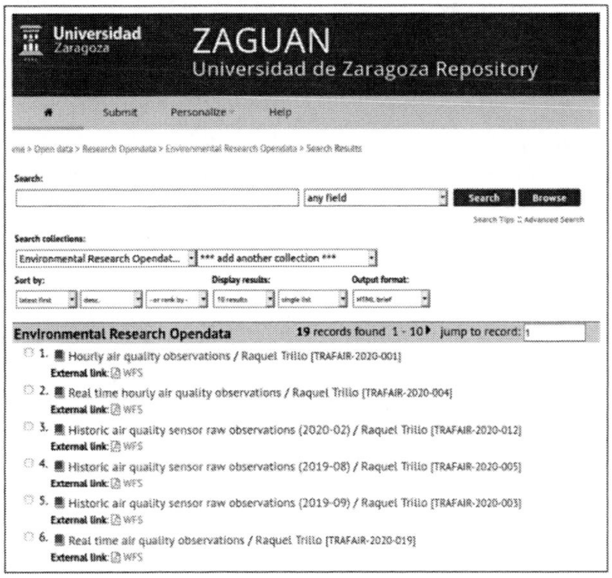


Figure 5. TRAFAIR datasets at Zaguan repository (University of Zaragoza)

an integrated approach to solve jointly three challenges: the automatic generation of metadata from the *GetCapabilities* responses of OGC services; the generation of DCAT-based metadata; and the synchronized publication of data and metadata.

## VI. Conclusion and Future Work

We have proposed the workflow for the publication of Open Spatial Data that can be customized to other projects dealing with spatial data that must be publicly accessible. Besides, we have demonstrated how GeoDCAT-AP metadata can be applied in a real use case to describe more specifically spatial data than other more general metadata vocabularies based on DCAT. The TRAFAIR project proposes a light adoption of GeoDCAT-AP that is feasible with minimum resources: all the proposed metadata elements are also included as part of the metadata elements included in DCAT-AP, and all these metadata elements can be edited through CKAN servers (either manually or through the CKAN API).

However, we must admit that not all GeoDCAT-AP guidelines to fill metadata elements could be followed by local Open Data portals because they must comply with constrained profiles of DCAT-AP imposed by national governments, which are beyond the control of project members. For instance, TRAFAIR proposes the use of INSPIRE data themes (i.e. "atmosphere" and "environment facilities") as values for *dcat:theme* because GeoDCAT-AP aims to be compliant with INSPIRE metadata rules, but unfortunately, the rules of our local portals force us to select a theme from a very limited controlled vocabulary established by the corresponding national government. A similar case occurs with metadata elements like *dct:provenance*, which is included both in DCAT-AP and GeoDCAT-AP but not in the Spanish subset of DCAT-AP [27].

As future work, we plan to integrate the software that we have developed for the automatic generation and publication of metadata as a new plugin of CKAN, or as an extension of existing *ckanext-spatial* plugin. Another work in progress is the evaluation of the quality of metadata according to several approaches like the Metadata Quality Assurance methodology [28] or the ISO 19157-based method for metadata quality analysis [29].

## Acknowledgment

This research has been supported by the TRAFAIR project 2017-EU-IA-0167, co-financed by the Connecting Europe Facility of the European Union. The contents of this publication are the sole responsibility of its authors and do not necessarily reflect the opinion of the European Union.

## References

[1] TRAFAIR consortium, "The website of TRAFAIR project (Understanding traffic flows to improve air quality)," 2020. [Online]. Available: http://trafair.eu/[retrieved:October,2020]

[2] L. Po et al., "TRAFAIR: understanding traffic flow to improve air quality," in 2019 IEEE International Smart Cities Conference, ISC2 2019, Casablanca, Morocco, October 14-17, 2019. IEEE, 2019, pp. 36–43. [Online]. Available: https://doi.org/10.1109/ISC246665.2019.9071661[retrieved:October,2020]

[3] Publications Office of the European Union, "The website of the European Data Portal," 2020. [Online]. Available: https://www.europeandataportal.eu/en[retrieved:October,2020]

[4] European Commission, "GeoDCAT Application profile for data portals in Europe, GeoDCAT-AP v1.0.1," 2016. [Online]. Available: https://joinup.ec.europa.eu/release/geodcat-ap/101[retrieved:October,2020]

[5] ——, "Commission Regulation (EC) No 1205/2008 of 3 December 2008 implementing Directive 2007/2/EC of the European Parliament and of the Council as regards metadata," European Union, Tech. Rep., 2008.

[6] CKAN Association, "The CKAN website," 2020. [Online]. Available: https://ckan.org/[retrieved:October,2020]

[7] International Organization for Standardization (ISO), "ISO 19115-1:2014. Geographic information - Metadata - Part 1: Fundamentals," Geneva, CH, Tech. Rep., 2014.

[8] W3C, "Data Catalog Vocabulary (DCAT) - Version 2. W3C Recommendation 04 February 2020," 2020. [Online]. Available: https://www.w3.org/TR/vocab-dcat/[retrieved:October,2020]

[9] European Commission, "DCAT Application Profile for data portals in Europe, DCAT-AP v2.0.0," 2019. [Online]. Available: https://joinup.ec.europa.eu/solution/dcat-application-profile-data-portals-europe/release/200[retrieved:October,2020]

[10] INSPIRE MIG, "Technical Guidelines for implementing dataset and service metadata based on ISO/TS 19139:2007," INSPIRE Maintenance and Implementation Group (MIG), INSPIRE Maintenance and Implementation Group (MIG). Version 2.0.1, 2017. [Online]. Available: http://inspire.ec.europa.eu/id/document/tg/metadata-iso19139[retrieved:October,2020]

[11] Open Source Geospatial Foundation, "API for GeoServer features," 2020. [Online]. Available: https://docs.geoserver.org/latest/en/api/#/latest/en/api/1.0.0/featuretypes.yaml[retrieved:October,2020]

[12] ——, "API for GeoServer coverages," 2020. [Online]. Available: https://docs.geoserver.org/latest/en/api/#/latest/en/api/1.0.0/coverages.yaml[retrieved:October,2020]

[13] D. Öttl et al., "Lagrangian dispersion modeling of vehicular emissions from a highway in complex terrain," Journal of the Air and Waste Management Association, vol. 53, 2003, pp. 1233–1240.

[14] T. Kralidis, "The OWSLib Python package," 2020. [Online]. Available: https://geopython.github.io/OWSLib/[retrieved:October,2020]

[15] CKAN Association, "The CKAN API guide," 2020. [Online]. Available: https://docs.ckan.org/en/2.8/api/index.html[retrieved:October,2020]

[16] Open Knowledge, "ckanext-dcat - RDF DCAT to CKAN dataset mapping," 2015. [Online]. Available: https://github.com/ckan/ckanext-dcat#rdf-dcat-to-ckan-dataset-mapping[retrieved:October,2020]

[17] Regione Emilia-Romagna, "The datasets of Comune di Modena at the Open Data website of Regione Emilia-Romagna," 2020. [Online]. Available: https://dati.emilia-romagna.it/dataset?organization=comune-di-modena[retrieved:October,2020]

[18] Concello de Santiago de Compostela, "The Open Data website at Concello de Santiago de Compostela," 2020. [Online]. Available: https://datos.santiagodecompostela.gal/es[retrieved:October,2020]

[19] Universidad de Zaragoza, "The environmental research open data at Zaguan (Universidad de Zaragoza Repository)," 2020. [Online]. Available: https://zaguan.unizar.es/collection/opendata-investigacion-medioambiental?ln=en[retrieved:October,2020]

[20] J. Nogueras-Iso et al., SDI Convergence: Research, Emerging Trends, and Critical Assessment. The Netherlands Geodetic Commission (NGC), 2009, ch. Development and deployment of a services catalog in compliance with the INSPIRE metadata implementing rules.

[21] Open Source Geospatial Foundation, "Catalog Services for the Web (CSW) - ISO Metadata Profile, GeoServer Community Module," 2020. [Online]. Available: https://docs.geoserver.org/stable/en/user/community/csw-iso/index.html[retrieved:October,2020]

[22] ——, "GeoNetwork User Manual v2.10.4-0, Harvesting OGC Services," 2020. [Online]. Available: https://geonetwork-opensource.org/manuals/2.10.4/eng/users/managing_metadata/harvesting/ogcwxs/index.html#ogcwxs-harvester[retrieved:October,2020]

[23] A. J. Florczyk, J. Nogueras-Iso, F. J. Zarazaga-Soria, and R. Béjar, "Identifying orthoimages in web map services," Computers & geosciences, vol. 47, 2012, pp. 130–142.

[24] J. Gil-Altaba, L. Díaz-Sánchez, C. Granell-Canut, and J. Huerta-Guijarro, "Open source based deployment of environmental data into geospatial information infrastructures," International Journal of Applied Geospatial Research, vol. 3, no. 2, 2012, p. 6–23.

[25] A. Perego, A. Friis-Christensen, and M. Lutz, "GeoDCAT-AP: Use cases and open issues," in Smart Descriptions & Smarter Vocabularies (SDSVoc) workshop. Amsterdam, 30 Nov - 1 Dec 2016. [Online]. Available: https://www.w3.org/2016/11/sdsvoc/SDSVoc16_paper_25[retrieved:October,2020]

[26] Open Knowledge, "ckanext-spatial - Geo related plugins for CKAN," 2015. [Online]. Available: https://docs.ckan.org/projects/ckanext-spatial/en/latest/[retrieved:October,2020]

[27] Ministerio de Hacienda y Administraciones Públicas, "Resolución de 19 de febrero de 2013, de la secretaría de estado de administraciones públicas, por la que se aprueba la norma técnica de interoperabilidad de reutilización de recursos de la información." Boletín Oficial del Estado, Lunes 4 de Marzo de 2013, 2013. [Online]. Available: http://www.boe.es/boe/dias/2013/03/04/pdfs/BOE-A-2013-2380.pdf[retrieved:October,2020]

[28] Publications Office of the European Union, "Metadata Quality Assessment Methodology. How EDP measures the quality of harvested metadata," 2020. [Online]. Available: https://www.europeandataportal.eu/mqa/methodology[retrieved:October,2020]

[29] M. Ureña-Cámara, J. Nogueras-Iso, J. Lacasta, and F. Ariza-López, "A method for checking the quality of geographic metadata based on iso 19157," International Journal of Geographical Information Science, vol. 33, no. 1, 2019, pp. 1–27.

# A Mobile Application to Share Georeferenced Tourist Experiences on a Discrete Global Grid

Rubén Béjar, Muhammad Umer and Javier Martínez-Fernández

Advanced Information Systems Laboratory (IAAA)
Aragon Institute for Engineering Research (IA3)
Universidad Zaragoza
c/ Mariano Esquillor s/n 50018, Zaragoza, Spain
Email: rbejar@unizar.es, m.umer@unizar.es, 737910@unizar.es

Jorge Dieste-Hernández, Ondřej Kratochvíl and Carlos López-Escolano

Study Group for Spatial Planning (GEOT)
Institute of Research into Environmental Sciences (IUCA)
Universidad Zaragoza
c/ Pedro Cerbuna, 12, 50009, Zaragoza, Spain
Email: jorgediestehernandez@gmail.com, ondrej@geogis.es, cle@unizar.es

***Abstract*—This work presents the prototype of a mobile application designed to make it easy for tourists to provide their opinions about the places they visit. Two characteristics make this application innovative. The first one is the use of a discrete global grid to collect the data, as discrete global grids are now just starting to be integrated with other Geographic Information System (GIS) technologies. And the second one is the strong emphasis on the emotions those places evoke on the users, an emphasis which is guided by the emotional cartography perspective. Besides this, tools are being added to allow the tourists to see and extend the views provided by previous visitors, in a collaborative, volunteered geographic information way.**

***Keywords–Emotional Cartography; Collaborative GIS; Discrete Global Grid System; DGGS; Tourism.***

## I. Introduction

The United Nations World Tourism Organization points out that tourism is being transformed, by means of digital technologies, in order to offer, among other things, "hyper-personalized customer experiences" [1]. Indeed, it is more and more common for tourists to value the possibility to enjoy, discover and share personal experiences, as shown for instance in the rise of platforms such as TripAdvisor or Expedia. With these platforms, tourists make decisions related to their choice of destination based on the experiences, opinions and judgement of others, and can then design personal experiences for themselves.

Edward W. Soja defines a conceived space as objective, qualifiable and mappable, and a perceived space as subjectively experimented, imagined and desired. And then he adds a third space, the lived one, as the summary of the other two [2]. Emotional cartography is a methodological process to represent the emotional spaces that form the territory [3]. The application of emotional cartography to crowdsourced tourist experiences, allows us to represent the emotional spaces that form the touristic places and analyze them along with the physical spaces.

This work discusses the prototype of a mobile web application which is being developed to capture georeferenced emotional data from the tourists that visit certain places. The analysis of these data within the paradigm of the emotional cartography will allow to make personalized recommendations for more personalized experiences such as destination branding, by associating certain places with the emotions that these places tend to elicit. The application will provide its users with the data created by previous visitors. They will be able to create new data, or to add their perceptions to areas created by previous users, in a collaborative GIS way.

The rest of the paper is organized as follows: in Section II, the related work section describes some relevant related research. Section III describes some technical and functional aspects of the prototype application being developed. Finally, Section IV summarizes the work done so far, and describes some expected future results.

## II. Related Work

Geographic Information Systems have been used to make tourism more interactive and user informed by making use of the location data. Some recent examples of this are [4], for nature-based tourism, and [5] for tourism marketing.

Location-aware mobile devices and specialized applications may provide tourists with valuable information to enjoy their trips and adapt them to their interests. A recent review of context-aware tourism applications, [6], analyzed them from four dimensions: knowledge acquisition, knowledge representation, knowledge processing and services offered to tourists. One of its conclusions is that the acquisition of the knowledge required to make these applications useful needs to take into consideration the crowdsourced feedback. This does not only need to include the location of the tourists; their experiences should be analyzed in a high spatial resolution to be studied in detail [7].

Those technologies that allow us to explore mental and emotional lives in non-invasive ways, and from a certain distance, are registering a fast acceleration [8]. With them, and through the use of emotional cartography, we can develop new geographical knowledge of places, such as those oriented towards tourism, and therefore new economic and social development. We can observe and analyze how the experiences in those spaces configure the life there, and we can get a deeper understanding of both their physical and human geographies

[9]. This in turn may contribute to discover new therapeutic possibilities of the places and to create a new image that makes them different from others. This kind of "branding" is essential for touristic development [10]. Destination management and marketing studies can also benefit from more research on emotions and tourism [11].

Collaborative mapping development, such as the OpenStreetMap (OSM), has become a trend in recent years. It uses community engagement to produce quality data and applications that may empower multiple sectors [12]. When compared to passive crowdsourcing, active collaborative GIS has been observed to generate more fine-scale data with a more flexible value range, which is better suited for management and analytics [13]. Collaborative GIS is currently been used, or at least proposed, in diverse domains such us landscape inventory creation [14], city models creations [15], and satellite imagery analysis [16].

Discrete Global Grid Systems (DGGS) are spatial information frameworks which divide the surface of the Earth in tessellations of discrete cells [17]. These cells are organized in a hierarchical fashion forming a multi-resolution grid. These grids are intended to be information grids, not navigation grids, and thus, issues such as quantization operations, i.e., assigning and retrieving data to/from cells, and algebraic operations on the cells and their contents must be defined by the different DGGSs. A DGGS must also provide a way to address, i.e., identify, each individual cell. There are proposals that build on that capability to address any area defined on a given DGGS [18]. The rHEALPix is a cubic geodesic DGGS, which is compatible with the OGC proposal [19]. Its cells, once projected onto the plane, are squares.

## III. The Prototype

Following the research goals, the application is designed to provide the users with an easy to use interface for capturing data. Data collection through the GUI is done by using the rHEALPix DGGS quadrilateral grids with the parameter $nside = 3$ (i.e., each square is divided into nine in the next resolution level). Only grids of resolution levels from 8 to 11 are shown (i.e., resolutions ranging from approximately 1.5 km, to around 55 meters). The users can navigate to their desired region using zoom in/out and panning functionality, and then select one or more grid cells just by tapping on their smartphone screen. The selected area will be added to the database with the submit button after fulfilling the information associated to it. The GUI of the current prototype is shown in Figure 1. Cells with a brighter color are those which have been selected by the user.

The source code of the prototype is currently available in two GitHub repositories: https://github.com/IAAA-Lab/grid-field and https://github.com/IAAA-Lab/grid-server. The first one includes the web application tier and the Django server with the grid models which are stored in PostgreSQL/PostGIS. The second one includes the Django Representational State Transfer (REST) based web server tier along with the rHEALPix and MongoDB components.

### A. Architecture

The main components of the application are shown in Figure 2. The application is entirely built with open source

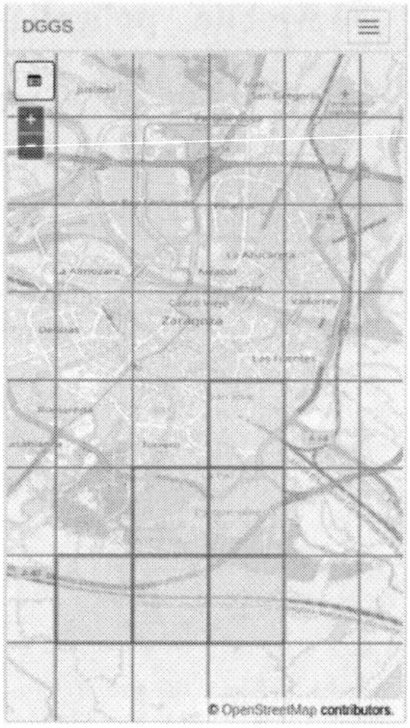


Figure 1. Application GUI: selection of an area based on a grid.

technology. We followed a three-tier architecture: web application, web server, and databases. In the web application tier, we focused on a mobile-first design as we expect this application to be used mainly with smartphones. The web server tier is built upon the Python Django framework, which provides the server side logical processing as well as the connectivity between the users and the database. The web application tier connects there through the Django REST framework Application Programming Interface (API).

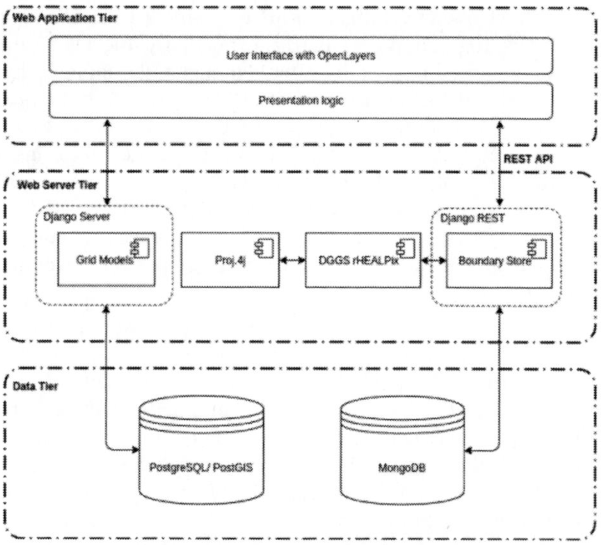


Figure 2. Component diagram: the 3-tier architecture of the application.

The web application tier, i.e., the frontend, uses the OpenLayers mapping library to display the rHEALPix grids on a base map, which currently is OpenStreetMap. The grids are retrieved in GeoJSON from the PostgreSQL database, with the PostGIS extension. For efficient processing, only the grids within the user extent are requested from the server and displayed.

Besides the access to the grid models stored in PostgreSQL, the web server tier also contains the DGGS-rHEALPix component. This component allows to store and retrieve sets of cells from rHEALPix that cover a given area, we call them Boundaries, which may be associated to some arbitrary JSON data. The storage and retrieval uses a MongoDB database to provide persistence. This rHEALPix grid component has been exposed as a Django REST framework API, and it also includes some basic level support for importing and exporting data from other GIS models and formats. This API is used by the frontend to manage the user generated geographic data based on the DGGS.

### B. Usage

The users of this application, i.e., the tourists, will not choose among existing, common geographic features or points of interest. As pointed out before, they will be defining areas on the geodetic grid displayed on the application by touching and selecting/deselecting cells over the base map. Then, they will add some information associated to that area, i.e., the emotions they felt when visiting those areas. We are allowing users to choose freely the areas they want because we are specifically interested in how tourists see and feel the space around them, i.e., Edward W. Soja perceived space, without imposing too many constraints on their choices.

However, the application will also be able to show the areas drawn by other tourists and the emotions they associated to those areas. In this way, the perceived space of some users might become the conceived space for others. Indeed, we expect that some areas will be seen as "natural limits" for certain features in the real space, and many tourists will reuse them, while other places should prove themselves to be more diffuse and difficult to delimit.

All these areas will be stored in the application as Boundaries associated to the emotional data. The users will provide the kind of emotion they are feeling by choosing among the universal emotions pointed out in the model by Ekman and Cordaro [20]: anger, fear, surprise, sadness, disgust and happiness (contempt is left out because it does not make sense to feel contempt about something which is not a person or group of people). It will be possible to choose more than one emotion, as it is possible for a person to feel different emotions about the same place, even at the same time. Besides this essential input, the application will allow to collect other data from the users, both automatic (, the date and time) and manual (e.g., a description of the place). This workflow is based on previous works where paper maps and other generic mobile GIS applications were used [21], and we expect to collect information that will allow us to produce similar results, like the emotional cartography shown in Figure 3 but in a more automatic way.

### C. The role of the DGGS

This prototype uses a DGGS to constrain the geometries of the areas that its users (the tourists) find of interest. This is driven by two main hypothesis. First of all, we hypothesize that it should be easier for non-expert users to draw geographic areas of interest on their smartphones by simply touching existing cells from a grid instead of drawing polygons as commonly seen in vector-based GIS data capture applications. We do not intend for this areas to have precisely delimited borders, and tools that allow to do that could prove themselves more difficult to use. And second, once we decide to use grids, it is rational to use some existing ones. The rHEALPix DGGS seems well suited to this project, as its cells are projected into squares. Although hexagons and triangles have some advantages, most non-expert users who have used any kind of map should be more familiar with rectangular grids.

Besides this, being intended to facilitate the integration of geographic data with different origins and scales, a DGGS should be a good candidate as a framework for the creation of emotional cartography. Within the frame provided by a DGGS, we can create a cartography where the objective is not to show, localize or collect all the details of a place, but the sensations and emotions that are disseminated over the space. And to do this in a diffuse, non-continuous way with different intensity, duration and temporality. The ability to select a resolution level, and a corresponding grid with specific cell sizes, from a DGGS provides us with a spatial framework that seems ideally suited to emotional cartography, where exactitude is not really possible and spatially diffuse areas are to be expected. For instance, it is perfectly reasonable to assume that a city in general sparks joy, but a particular neighborhood evoques sadness and a certain square in the neighborhood provides hope. The hierarchical spatial framework provided by a DGGS should facilitate this kind of multi-scale data collection.

## IV. Conclusion and Future Work

This paper has described the prototype of a mobile web application intended to capture georeferenced emotional data from tourists. This application uses a DGGS as a framework both to capture and analyze the data. We hypothesize that this kind of grid-based framework makes it easier for the users to delimit the diffuse geographic areas that can be associated to the different emotions they felt over the place. We also consider that the hierarchical nature of the grids in a DGGS will facilitate the analysis of data at different scales as needed.

Once we have deployed an operative version of the application, we intend to validate, or refute, the hypothesis that have driven the design of the application. Besides testing the application itself, the data collected will be analyzed, and cartographically presented, under the paradigm of the emotional cartography.

After that, we want to advance on the collaborative part of the application. As different users provide their own views about a territory, we should provide them with to opportunity to create new data, or to add their perceptions to areas created by previous users, in a collaborative GIS way. It will be necessary to develop a system where the mechanisms implemented to solve conflicts, merge similar entries and find possible relations allow to study how consensus and compromises emerge, or not.

Finally, we also expect to find cartographic challenges to portray a collaborative work based on personal feelings, which are not only different for different people, and associated to slightly, or not so slightly, different locations, but that may vary for example under a different weather, or just with the time of the day.

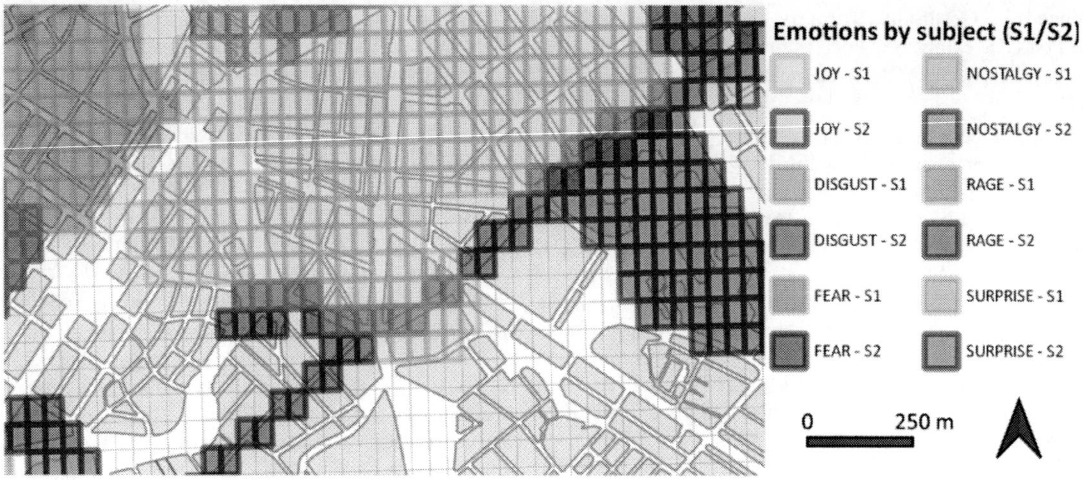


Figure 3. An example of emotional cartography using a geodesic grid to depict emotions associated to areas on a base urban map.

## ACKNOWLEDGMENT

This work has been partially supported in 2019-2020 by the project LMP19_18, with Aragón FEDER funds 2014-2020, "Construyendo Europa desde Aragón", and by the project T59_20R of the Aragón Government.

## REFERENCES

[1] UNWTO, Ed., International Tourism Highlights, 2019 Edition. United Nations World Tourism Organization, 2019, [retrieved: October, 2020]. [Online]. Available: https://www.e-unwto.org/doi/pdf/10.18111/9789284421152

[2] E. W. Soja, Thirdspace: Journeys to Los Angeles and other Real-and-Imagined Places. Oxford: Blackwell, 1996.

[3] Raqs Media Collective, M. van de Drift, S. B. Davis, R. van Kranenburg, S. Hope, and T. Stafford, Emotional Cartography - Technologies of the Self, C. Nold, Ed., 2009, [retrieved: October, 2020]. [Online]. Available: http://emotionalcartography.net/

[4] O. Ghorbanzadeh, S. Pourmoradian, T. Blaschke, and B. Feizizadeh, "Mapping potential nature-based tourism areas by applying GIS-decision making systems in East Azerbaijan Province, Iran," Journal of Ecotourism, vol. 18, no. 3, 2019, pp. 261–283.

[5] H. Albuquerque, C. Costa, and F. Martins, "The use of Geographical Information Systems for Tourism Marketing purposes in Aveiro region (Portugal)," Tourism Management Perspectives, vol. 26, 2017, pp. 172–178.

[6] F. Leal, B. Malheiro, and J. C.Burguill, "Context-aware tourism technologies," The Knowledge Engineering Review, vol. 33, no. e13, 2018, pp. 1–26.

[7] N. Shoval and A. Birenboim, "Customization and augmentation of experiences through mobile technologies: A paradigm shift in the analysis of destination competitiveness," Tourism Economics, vol. 25, no. 5, 2019, pp. 661–669.

[8] C. Ellard, Places of the Heart: The Psychogeography of Everyday Life. Oxford: Bellevue Literary Press, 2015.

[9] E. Olmedo, "Cartographie sensible: tracer une géographie du vécu par la recherche-création (Sensitive cartography: tracing a geography of lived experience through research-creation)," Thèse de doctorat en Géographie, École doctorale de Géographie de Paris, Paris, France, November 2015.

[10] J. Nogué and J. Vela, "Geographies of affect: In search of the emotional dimension of place branding," Communication and Society, vol. 31, 2018, pp. 27–44.

[11] A. Scuttari and H. Pechlaner, Emotions in Tourism: From Consumer Behavior to Destination Management. Springer International Publishing, 2017, pp. 41–53.

[12] J. Panek and R. Netek, "Collaborative mapping and digital participation: A tool for local empowerment in developing countries," Information, vol. 10, 2019, p. 255.

[13] L. Muñoz, V. H. Hausner, C. Runge, G. Brown, and R. Daigle, "Using crowdsourced spatial data from Flickr vs. PPGIS for understanding nature's contribution to people in Southern Norway," People and Nature, vol. 2, 2020, pp. 437–449.

[14] I. Santé et al., "The Landscape Inventory of Galicia (NW Spain): GIS-web and public participation for landscape planning," Landscape Research, vol. 44, 2018, pp. 212–240.

[15] I. Prieto, J. L. Izkara, and R. Béjar, "A continuous deployment-based approach for the collaborative creation, maintenance, testing and deployment of CityGML models," International Journal of Geographical Information Science, vol. 32, no. 2, 2018, pp. 282–301.

[16] W. Su, D. Sui, and X. Zhang, "Satellite image analysis using crowd-sourcing data for collaborative mapping: current and opportunities," International Journal of Digital Earth, vol. 13, 2018, pp. 1–16.

[17] M. Purss, Ed., The OpenGIS Abstract Specification - Topic 21: Discrete Global Grid Systems Abstract Specification). Open Geospatial Consortium, August 2017, no. OGC 15-104r5.

[18] R. Béjar, M. Á. Latre, F. J. López-Pellicer, J. Nogueras-Iso, and F. J. Zarazaga-Soria, "On the problem of providing unique identifiers for areas with any shape on discrete global grid systems," in Accepted Short Papers and Posters from the 22nd AGILE Conference on Geo-information Science, Limassol, Cyprus, 17-20 June, 2019, [retrieved: October, 2020]. [Online]. Available: https://agile-online.org/images/conference_2019/documents/short_papers/58_Upload_your_PDF_file.pdf

[19] R. G. Gibb, "The rHEALPix Discrete Global Grid System," in IOP Conference Series: Earth and Environmental Science, ser. IOP Conference Series: Earth and Environmental Science, vol. 34, Apr. 2016, p. 012012.

[20] P. Ekman and D. Cordaro, "What is meant by calling emotions basic," Emotion Review, vol. 3, no. 4, 2011, pp. 364–370.

[21] J. Dieste, O. Kratochvíl, M. P. Serrano, and A. Pueyo, "Los mapas emocionales: Un instrumento para la mejora del conocimiento de los espacios metropolitanos (Emotional maps: a tool for improving knowledge of metropolitan areas)," in Libro de Actas del XXXVI Congreso de la Asociación de Geografía Española. Crisis y espacios de oportunidad. Retos para la Geografía, Valencia, Spain, 2019, pp. 1523–1524, [retrieved: October, 2020]. [Online]. Available: https://www.age-geografia.es/site/wp-content/uploads/2020/01/Actas-Congreso-Conclusiones-AGE-VLC2019_compressed_reduce-1.pdf

# Temporal Distance Map: A Warped Isochrone Map Depicting Accurate Travel Times

Elijah Nacar, Devak Nanda
Texas Academy of Mathematics and Science,
University of North Texas
Denton, USA
e-mail: elijahnacar@my.unt.edu
e-mail: devaknanda@my.unt.edu

Blake Albert, Christian Panici
Department of Computer Science,
Loyola University
Chicago, USA
e-mail: oli@oleacapita.com,
e-mail cpanici@luc.edu

Mark V. Albert
Computer Science and Computer Engineering,
University of North Texas
Denton, USA
e-mail: Mark.Albert@unt.edu

***Abstract* - The presented Temporal Distance Mapping tool creates a visual representation in which distance on the map represents travel time rather than physical distance. In the age of routing applications, most people are more concerned with the time it will take to reach a destination rather than its physical distance. A river, mountain, or even traffic can make nearby points on a map seem distant by comparison, while highways and fast public transportation lines can seem to bring distant physical locations together. Utilizing travel data, we can morph the shape of any mapped region to accurately depict travel time. First, a traditional static image map for a specific location of interest is overlaid with a grid of points. The travel time from the center point to each grid point is calculated. Each grid point is then shifted radially toward or away from the center point depending on calculated travel time. The rest of the map pixels are then shifted according to the new gridpoint locations using an affine transformation. In this way, the original map is warped to represent travel time relative to the center point. Although two destinations on a traditional map may have the same physical distance, the travel time may be orders of magnitude different due to barriers or access to public transportation. This map better represents access to the surrounding area, and also provides a compelling visual representation to understand the local community in the context of what matters most to them - their time.**

***Keywords - euclidean; temporal distance; isochrone; metadata.***

## I. INTRODUCTION

Travel time is one of the main concerns people have when viewing a map, however, travel time is poorly represented on most physical maps. Transportation systems bring distant physical points in closer proximity in terms of time, while physical barriers can dramatically impact the travel time between closely located points on a map [1]. When faced with the prospect of visualizing travel time, most turn to the isochrone map [2][3]. An isochrone map uses contour lines to represent equivalent travel time from a single location [4]. The isochrone map is used in geographic [5][6], clinical [7], and astrophysical research [8]. Isochrone maps are widely used in research, however, they are also limited in public use [9]. Isochrone maps are not a common tool in the general populace given the complexity to reach them with overlaid contours. It would be more beneficial to have a map representation which directly depicts the travel time for the everyday user.

In the age of vehicle routing applications [10], drivers are much more interested in how long it takes to reach a location rather than the physical distance shown on most maps. Isochrones contours display these "temporal distances" but are only an overlay on a representation that is less directly relevant to a person's experience of the world around them. The time it takes a person to reach a destination is far more important than the physical distance, particularly in cities and would best be represented directly in the underlying representation, rather than as an overlay on a less relevant depiction of physical distance.

We created a tool that uses the information present in a polar isochronic map [11] and morphs the static image to present travel time as the distances. Web mapping services [12] currently overlay alternate routes and travel time information onto a traditional map representation. This approach only provides information for travel time between two locations. A user may want to have a better understanding of their surrounding area to consider alternate destinations. Using current commercial mapping tools, if someone was interested in comparing travel times within an area, they would need to calculate the travel distance between every single point of interest [13]. We offer a tool that expedites this process while also offering visual clarity that will better allow a comparison of travel times at a glance.

The remainder of this paper is organized as follows: Section II explores our method for warping the images based on calculated travel time. Section III demonstrates application of the temporal distance map tool for a variety of locations. Finally, the paper concludes with Section IV.

All information and documentation relevant to the research can be found on the github repository [14].

## II. METHODS

As an overview, the method for generating the warped map begins with the creation of an overlaid and spaced rectangular grid of points. The travel time to the grid points is calculated, and the grid points are moved radially in proportion to travel time. The pixels of the original map are then transformed based on the new locations of the grid points. We will now step through the process in more detail.

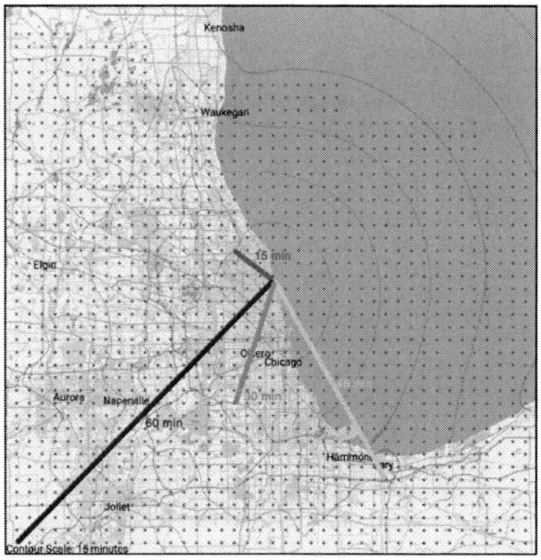


Figure 1. Plot of 961 euclidean points overlaid the static image of the location. Arrows and times represent the travel time between each corresponding geographic coordinate and the center

To begin, we used Bing Maps API [14] calls to get a standard static image of the location. Then, on that image, we overlaid a mesh of 961 points. The corresponding latitude and longitude were mapped to a normalized coordinate system ranging from 0 to 1 along each dimension (Figure 1). For instance, (0.5, 0.5) is the center point on the map, and if the image depicted a latitude range of +56 to +60 (or 56 N to 60 N) and a longitude range of +40 to +44 (or 40 E to 44 E) then the central point would be the middle of each range or the geographic coordinate (58 N, 42 E).

Utilizing the Bing Maps Distance Matrix, we were able to efficiently find the travel time between every geographic coordinate and the central location (Figure 1). Then, we proceeded to remap the grid coordinates based on the calculated travel time. First, we calculated an estimate of one minute of travel time in the normalized coordinates by dividing the euclidean distance of the temporally furthest point on the map by its travel time. Using this estimate, we then calculated the new distance the grid point should be from the center. We maintained the same angle of the grid point relative to the center. This information provided a radius and angle from the center point for the new transformed coordinate. Effectively this shifted each grid point radially from the center point depending on calculated travel time.

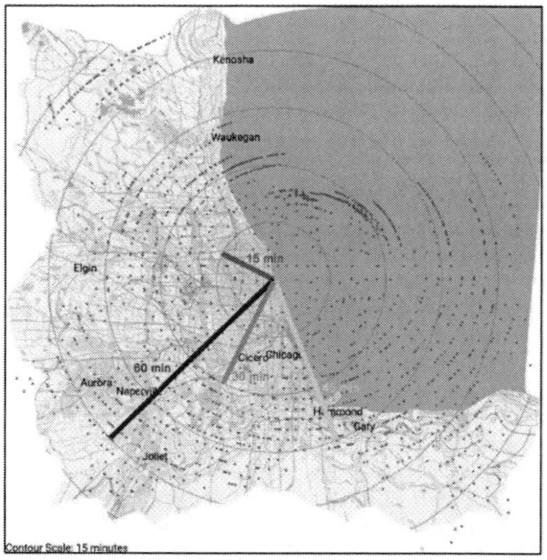


Figure 2. The plot of 961 euclidean points post-transformation based on travel time. The points now form concentric circles around the center and the travel time is now properly represented and uniform around the image.

Each grid point now has original coordinates and transformed coordinates; however, they were spaced out significantly given the computational resources necessary to calculate travel time between all grid points and the center point. In order to visualize the new coordinates, the pixel coordinates of the original map also had to be transformed. Each pixel of the original image was converted to normalized coordinates, and an affine transformation based on the surrounding grid points was performed to transform the original, normalized pixel coordinates into the new map coordinates (Figure 2).

Additionally, this coordinate transform was also used to create an animation to more readily observe the effects of warping due to travel time. The old and new coordinates were then linearly interpolated from time t=0 at the original coordinates to t=1 for the new transformed coordinates.

## III. RESULTS AND DISCUSSION

In this section, we display several applications of the Temporal Distance Map to observe and discuss the effect. The demonstrations are for Pennsport, Pennsylvania; Miami, Florida; and Kansas City, Kansas. In these map transformations, notable changes can be seen due to distinct geographic features and infrastructure.

### *A. Transformation of Pennsport and the Delaware River*

The Delaware River acts as a natural boundary between Pennsylvania and New Jersey. Pennsport, a city on the edge of the river with close proximity to both the Walt Whitman and Benjamin Franklin Bridges, is depicted in Figures 3 and 4.

Figure 3. Static image of Pennsport from the Bing Maps API.

Figure 4. Morphed version of Pennsport. The location is transformed based upon travel time from the central(red) point.

After transforming the image, the area immediately around Pennsport begins to stretch relative to other areas outside the first 15 minute contour. This is due to the availability of ways to cross the river. Someone situated in the heart of Pennsport needs to drive either North or South to one of the bridges in order to reach New Jersey on the opposite side; therefore, increasing the relative travel time. Notice how once the river is crossed (generally around the 15 minute contour) other areas begin to squeeze together due to the availability of roads once they cross the bridge.

### *B. Transformation of Miami Bay, Key Biscayne, and the Everglades/Francis Wildlife Management Area*

The Eastern Coast of Miami has a variety of islands, harbours, and keys. Key Biscayne is located to the South-East of Downtown Miami and is used as the center in Figure 5. Notably, Miami travel is more efficient along the coast given the geographic barriers to travel. Given these geographic barriers, the North-West and South-East corners of the map both stretch due to how long travel takes compared to travel along the coastal highways. The North-West corner is faced with a journey across the management area of both the Everglades National Park and the Big Cypress National Preserve, both of which have limited vehicle infrastructure. The South-East corner is blocked by Key Biscayne, any traveler that wants to reach that corner of the map will need to choose to drive around Key Biscayne on a boat or drive along Key Biscayne in its entirety, increasing travel time when compared to other regions on the map. There is, notably, a direct path through a gap in Key Biscayne that would allow a boater to travel to the South-East corner uninterrupted, thus resulting in the large uninterrupted zone of generally constant contour rings.

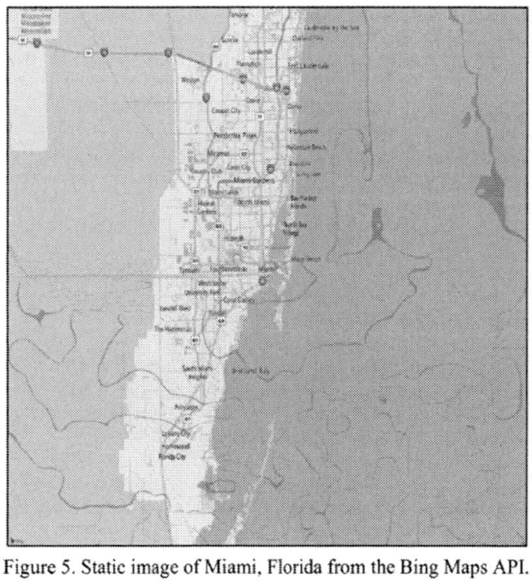

Figure 5. Static image of Miami, Florida from the Bing Maps API.

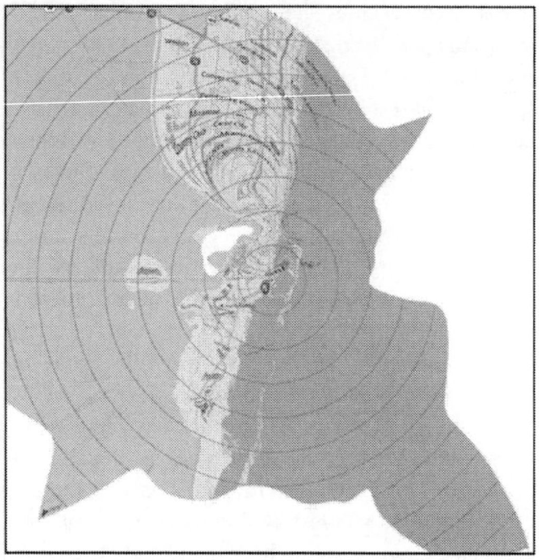

Figure 6. Morphed version of Miami, notice the warping of the ocean surrounding the Key Biscayne.

Notably, the acceptable methods of travel would alter the representations, so by altering the route options available (e.g., allowing use of tolls, ferries, etc.) the resulting representation would change.

*C. Transformation of Kansas City and Road Infrastructure*

Kansas City was chosen due to its transportation infrastructure. The city is completely covered by infrastructure, like highways and public transport systems, all of which influence the transformation depicted in Figure 5.

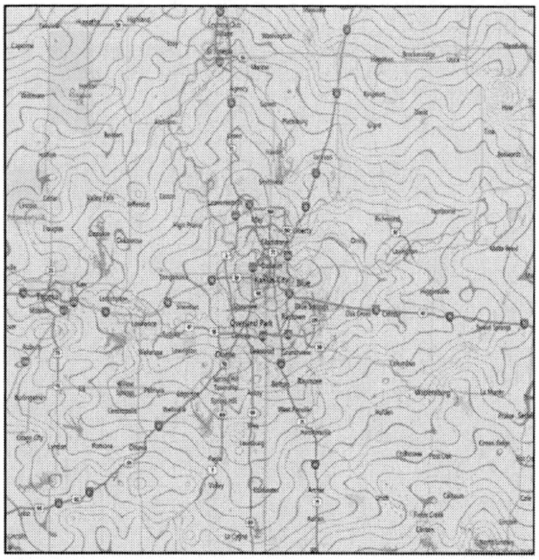

Figure 7. Static image of Kansas City, Kansas from the Bing Maps API.

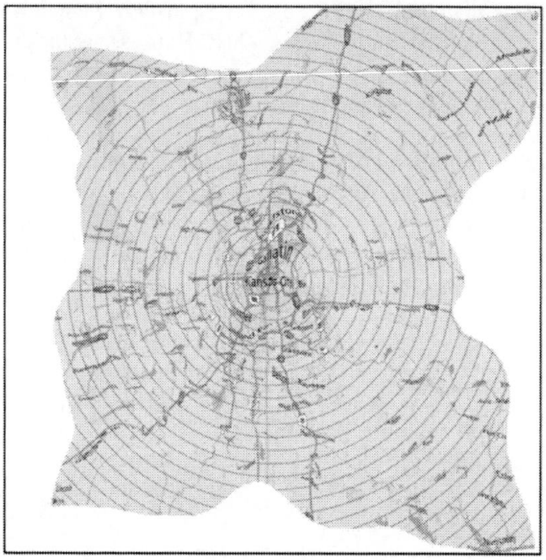

Figure 8. Morphed version of Kansas City, notice how the abundance of highways morph the region.

For the transformation in Figure 7, the contours reflect 5 minute intervals instead of the typical 15 minutes in other figures. This way, it is readily apparent how highways begin to shift the image of the map. First, highways allow a driver to cover a large distance in a relatively short period of time. Thus, the warped image contracts along highways, a pattern that is visible in any of the projections featuring a major roadway Additionally, while Kansas City is known for its widespread road infrastructure, the top right corner of the map is less connected with the city center. Therefore, during the transformation the map contracts everywhere except the top right corner, which is unreachable along a highway.

These examples represent a cross-section of geographic impacts on travel time. The river of Pennsport, the ocean and everglades of Miami, and the road infrastructure of Kansas City all warp the map in ways that are consistent with a local understanding of travel time, but depict that information visually in a way that is more direct to a casual observer.

## IV. CONCLUSION AND FUTURE WORK

This project was prompted by a desire to visualize access to nearby locations in a way that is more relevant to personal experience - travel time rather than physical distance. We created an application that allows a user to enter a geographic location and returns a transformed map of the region in a way that depicts travel time by distance on the map.

The transformations more directly represent the impact of geographic and infrastructure features on a person's experience navigating the local area. This new way of representing distance can be used to inform personal travel decisions by providing a more direct comparison between travel time and distance.

The new projection would be most useful for travelling in areas with unique geography that is unfamiliar to the traveller. By being able to readily compare alternate destinations relative to one's current location, travel time can be more intuitively used in selecting among the alternate destinations. The implementation uses readily available map API information, and a series of linear transformations and interpolations allow for commercial scalability.

There are a number of further advances possible. In this approach, locations were shifted radially from the center, however allowing for angular movement of points may have led to fewer artifacts in the warped representation. Additionally, this approach identified a center point, however, it is conceivable to create a representation without an arbitrary center. By observing travels times between all pairs of grid points, and creating a networked representation of grid points with connections weighted by travel time in a force-directed graph layout. This would create a representation stretching the image in slow-moving areas and compressing along fast corridors, but without identifying a single central point. This would enable a representation for an entire region that could be shared or marketed for everyone in the region.

The Temporal Distance Map presented here provides a new way to visualize travel time. Isochrone maps provide similar information, but this tool takes the concept a step further by morphing the underlying representation to make the information present in isochrone contours more directly accessible. This way the intuitive understanding of travel time for surrounding locations that a native resident feels is more accessible to people new to an area and more directly represents information of importance to them - time rather than distance.

## REFERENCES


[1] E. Bielecka and A. Bober, "Reliability analysis of interpolation methods in travel time maps - The case of Warsaw," Geodetski Vestnik, pp. 299-312, 2013

[2] J. van den Berg, B. Köbben, S. van der Drift, and L. Wismans, "Towards a Dynamic Isochrone Map: Adding Spatiotemporal Traffic and Population Data. Progress in Location Based Services 2018," Springer International Publishing, 2018. pp. 195–209.

[3] S. Bies and M. van Kreveld, "Time-Space Maps from Triangulations. Graph Drawing," Springer Berlin Heidelberg, pp. 511-516, 2013.

[4] R. A. Bryson, W. M. Wendland, J. D. Ives, and J. T. Andrews, "Radiocarbon Isochrones on the Disintegration of the Laurentide Ice Sheet," vol. 1, pp. 1–13, 1969.

[5]. A. Efentakis, N. Grivas, G. Lamprianidis, G. Magenschab, and D. Pfoser, "Isochrones, traffic and DEMOgraphics," Proceedings of the 21st ACM SIGSPATIAL International Conference on Advances in Geographic Information Systems. New York, NY, USA: Association for Computing Machinery, pp. 548–551, 2013.

[6] A. K. Darvishan, S. H. Sadeghi, and L. Gholami, "Efficacy of Time-Area Method in simulating temporal variation of sediment yield in Chehelgazi watershed, Iran," Annals of Warsaw University of Life Sciences, 2010.

[7] H. S. Oster, B. Taccardi, R. L. Lux, P. R. Ershler, and Y. Rudy, "Noninvasive Electrocardiographic Imaging," Circulation, vol. 96 pp. 1012–1024, 1997.

[8] D. A. VandenBerg, P. A. Bergbusch, and P. D. Dowler, "The Victoria-Regina Stellar Models: Evolutionary Tracks and Isochrones for a Wide Range in Mass and Metallicity that Allow for Empirically Constrained Amounts of Convective Core Overshooting," Astrophys J, 2006.

[9] N. Street, "TimeContours: Using isochrone visualisation to describe transport network travel cost," Final Report, Jun. 2006.

[10] P. Toth and D. Vigo, Vehicle routing: problems, methods, and applications. 2014.

[11] H. Sutanto, "Polar coordinate-based isochrone generation," US Patent. 6668226, 2003.

[12] D. Zhang et al., "Efficient evaluation of shortest travel-time path queries through spatial mashups," Geoinformatica, vol. 22, pp. 3–28, 2018.

[13] A. Ozimek and D. Miles, "Stata Utilities for Geocoding and Generating Travel Time and Travel Distance Information," Stata J, vol. 11, pp. 106–119, 2011.

[14] Microsoft Corporation, 2020. "Bing Maps Documentation - Bing Maps," [online] Docs.microsoft.com, Available at: <https://docs.microsoft.com/en-us/bingmaps/> [Accessed 17 June 2020].

# Identifying the Existence of Grass Coverage in Vineyards by Applying Time Series Analysis in Sentinel-2 Bands

Daniel A. Basterrechea[1], Lorena Parra[1,2], Jaime Lloret[1], and Pedro V. Mauri[2]
[1]Instituto de Investigación para la Gestión Integrada de zonas Costeras. Universitat Politècnica de València, Valencia, Spain
[2] Instituto Madrileño de Investigación y Desarrollo Rural, Agrario y Alimentario, Madrid, Spain
Email:dabasche@epsg.upv.es, loparbo@doctor.upv.es, jlloret@dcom.upv.es, pedro.mauri@madrid.org

*Abstract*— **The increasing tendency of the population puts pressure on vineyard farmers to supply food. In this context, grass coverage will be a low-cost solution to reduce outlays and reduce the maintenance of crops. In this paper, we present a remote sensing technique for determining the existence or absence of grass coverage in vineyards. To perform this study, we use Sentinel-2 images using red, green, and blue bands, as well as a water vapour band, near-infrared band, and normalized difference vegetation band. This technique has certain limits, such as low spatial resolution, and cloud presence when the images are obtained. The selected images have 10 m x 10 m spatial resolution, except for the band of water vapour band (60 m x 60 m). In this study, we propose the use of time-series analysis to overcome the problem of low spatial resolution. To perform this study, we obtain images from January and June of 2020. Using ArcGIS software, we applied different tools to obtain qualitative and quantitative information. Then, we analyzed the significance of observed differences in the time series analysis, applying Single Analysis of Variance. Our results indicate that the best results are related to the Near-infrared band with a p-value of 0.0020. Finally, the pixel values obtained for the time series analysis of Near-Infrared band indicate that the plots with grass coverage have values from -1000 to -1200. Meanwhile, values from -1200 to -1500 are found in the plots without grass coverage.**

***Keywords- Precision agriculture; Image processing; Sentinel-2 bands; Vineyard monitoring.***

## I. INTRODUCTION

The world population is increasing, expecting to reach between 8.1 billion and 10.6 billion by 2050. This growth causes the necessity to maximize the production of food [1]. The existing farming systems are not able to attain efficient food production levels. In this context, the research for new techniques is one of the concerns of the population instead of optimizing the use of natural resources. On the other hand, the pressure of the increment of food demand and the decrease in prices forces the farmers to try to increment their crop production. Thus, in some cases, farmers are using dangerous substances for the environment, massive use of water, or excess use of fertilizer to maximize their harvest. Nonetheless, the implementation of sustainable agriculture might increase productivity, minimizing the environmental impact of the activity. The proper management of natural resources will be an interesting way of solving the actual problem [3].

In this context, the inclusion of new technologies in Precision Agriculture (PA) has been revealed as a technological solution. PA consists of the application of different techniques for the management of the agricultural stock. Therefore, it is possible to obtain optimum control of the production and of the required resources, and guarantee the sustainability of the activity. PA can be a solution to the problem of food security [4]. Another alternative is the application of conservation agriculture, which is based on the minimum disturbance of soil and the maintenance of grass coverage in the crops [5]. Its main advantages are the reduction of erosion and the improvement of water retention. In order to evaluate the degree of adoption of this practice, we need to evaluate with remote sensing the existence of grass coverage.

One of the techniques used for monitoring crops is image processing. This technique has been used to a huge range of purposes such as detecting weed plants [6], monitoring the plant health [7], quantifying the harvest [8], and evaluating pollutant presence. To maximize the results, this method can be applied combined with terrestrial techniques [9]. One of the problems with image processing and remote sensing techniques is the low resolution of the used images. The images with high spatial resolution, are more expensive and are not usually utilized for general research. When the available spatial resolution cannot fit the requirements, there is an alternative, the use of time series analysis. The use of time series analysis has been done in [10] with sunflowers and Unmanned Aerial Vehicles (UAV).

In this paper, we propose the use of images from the Copernicus Sentinel-2 satellite, with 10 m pixel resolution, to determine the existence or absence of grass coverage in vineyards. This evaluation will be done using images obtained at different moments of the year. We apply this methodology in the inner region of Spain, in the crops of IMIDRA, which is located in Alcala de Henares (Madrid). To obtain the required accuracy, we include in this analysis the following information from the satellite, including red band, green band, blue band, Water Vapour Band (WVP), Near-Infrared Band (NIR), and Normalized Differential Vegetation Index (NDVI) which is a combination of the Red band and NIR band.

The rest of the paper is structured as follows. The related work is outlined in Section 2. Section 3 presents the different techniques and the process of the study. The results are discussed in Section 4. Finally, the conclusions and future work are summarized in Section 5.

## II. RELATED WORK

In this section, we outline the state of the art. In the summarized contributions, we include systems proposed for

PA and image processing techniques for monitoring different parameters of the land.

Sun et al. [11] used a multispectral image to monitor the chlorophyll content in the field. They applied different fertilizers to the crops and used a multispectral Charge-Coupled Device (CCD) camera to collect ground-based images in the green, red, and NIR bands. They developed a new Normalized Difference Vegetation Index (NDVI). Besides, they obtained the correlation between image parameters and chlorophyll content obtaining R2 of 0.88. They concluded that vegetation indices derived from a multispectral image could be used to monitor the chlorophyll content. Mishra et al. [12] applied advances in Object-Based Image Analysis (OBIA) and machine learning algorithms in dry savanna ecosystems for forest detection. To do this, they use remote sensing-based in the characterization of vegetation properties is savannas. In this case, they used a stack of Landsat Thematic Mapper (TM) imagery, NDVI, and topographic variables with six different scale factors resulting in a hierarchical network of image objects. Additionally, individual vegetation morphology classes differed in the segmentation scale at which they achieved the highest classification accuracy, reflecting their unique ecology and physiognomic composition. Finally, their results showed the utility of the OBIA.

Other authors, as Parra et al. [13] proposed image techniques to detect prejudicial weeds in lawns. To perform this study, they used a mathematical operation where the red, green, and blue bands, as well as, edge detection techniques, are used. Besides, they use a post-processing operation to reduce the false positives, changing the combination between the selected bands. Clever at al. [14] used the Sentinel-2 and Sentinel-3 images for the estimation of total crop and grass chlorophyll and N content by studying in situ crop variables and spectroradiometer measurements obtained for four different test sites. The obtained results confirmed the importance of the red-edge bands, particularly in Sentinel-2 for agricultural applications, because of the combination with its high spatial resolution of 20 m. Rokhmana et al. [15] displayed some practical experiences of using UAVs based platform for remote sensing in supporting PA mapping. They proposed a system based on the aerial platform from Radio-Controlled plane, point and shoots digital cameras, and data processing with digital photogrammetric mapping.

In this paper, we present a low-cost method for determining the existence of a grass coverage in the vineyard using plots with grass and others without grass coverage. To perform the study, we use Sentinel-2 images from different timelines. This application will be useful, to elaborate maps and analysis about the adoption of conservation agriculture. Furthermore, it can be used to study the need for specific actions to maximize its adoption in certain regions or to evaluate the changes in the agroecosystems after certain activities.

## III. MATERIALS AND METHODS

In this section, the used materials and the methodology for the analysis of the data are presented.

### A. *Selection of Satellite*

Satellite images are selected for detecting the grass coverage in vineyards. We use free satellite images to obtain a low-cost system of determining the existence of grass coverage. In this context, we decide to use images from Copernicus Sentinel-2. The Sentinel-2 is characterized by two polar-orbiting satellites placed in the same sun-synchronous orbit, phased at 180° to each other. Moreover, Sentinel-2 provides information every ten days, which is enough for our objective. Besides, this satellite has two types of images. Firstly, we have "Level-1C" products, which give information on the top of atmosphere reflectance in cartographic geometry, and "Level- 2A" offers data of the bottom of the atmosphere reflectance in cartographic geometry.

Moreover, Sentinel has a huge range of multispectral images, where the highest resolution is 10m x 10m. In this paper, we use only the six different bands included in Table 1 and the Normalized difference vegetation index (NDVI). Table 1 displays the characteristics of the used bands.

### B. *A proposed approach for the time series analysis*

Following we detail the principle that we follow to perform the time series analysis and detail the changes along the year in the studied plots with and without grass coverage.

Regarding the obtained images with Sentinel-2, each pixel contains information about the surface, which includes the vineyard, the soil, and if it grass coverage exists. Nonetheless, the grass coverage is not present during the entire year, its presence in maximum in winter and almost null in summer, see Figure 1.

In winter, the plant cover is greater due to the climate conditions. It is characterized by being wet and cold with a considerable rate of precipitation. On the other hand, in summer high temperatures predominate, where rainfall is drastically reduced, causing vegetation to wither and tend to disappear. These changes based on the season could be used to detect different pixel values in the bands. In this context, we hypothesize that, during the period in which grass coverage has a maximum presence (winter), the pixel values will be different from when the grass coverage is not present (summer) for plots with grass coverage.

TABLE I. SENTINEL-2 SPECTRAL BANDS

| Bands | Wavelength (nm) | Resolution (m) | Description |
|---|---|---|---|
| B2 | 490 | 10 | Blue |
| B3 | 560 | 10 | Green |
| B4 | 665 | 10 | Red |
| B8 | 842 | 10 | Visible and Near Infrared (VNIR) |
| B9 | 945 | 60 | Water vapour |

Nonetheless, for plots without coverage, the pixel values of both seasons will be very similar. The differences between the data from winter and summer based on our hypothesis are displayed in Table 2.

To perform the study, we select images from January and June. We select these images taking into account the live cycle of the vineyard and grass in the different parts of the seasons. This information is essential to select the crucial moments in which images are gathered. In this case, in January the vineyard does not have any leaves because it is the part of the year when the tree is pruned. On the contrary, in June the vineyard begins to have leaves. Meanwhile, grass coverage changes between winter and summer. In this area, winter is cold with a high precipitation rate, and summer is characterized by high temperatures of low precipitation.

### C. *Studied Zone*

The studied zone that we selected is located in the community of Madrid in the facilities of IMIDRA. We selected this location because there are huge vineyards where we have plots with grass coverage and others with non-coverage. It constitutes an optimum scenario to test the proposed system for monitoring the changes in the grass to determine the presence or absence of grass coverage.

We classify the selected plots in two types: The ones that contain grass coverage (GC=1), and the ones that do not present grass coverage (GC=0). The plots and the classification can be seen in Figure 2. We select seven plots, 4 of them without grass coverage, and 3 with grass coverage. We represent in red colour and label them as 1, the crops without grass coverage. On the other hand, we use the blue colour to represent plots with grass coverage, which are labelled as 0. In this context, we have the information on which plots present grass coverage and which do not.

### D. *The software selected of analysis of time series*

For the analysis, we use a combination of the different bands detailed in Table 2, from the different seasons. A specialized program is needed for treating the obtained satellite images. In this case, we select the ArcGIS [16].

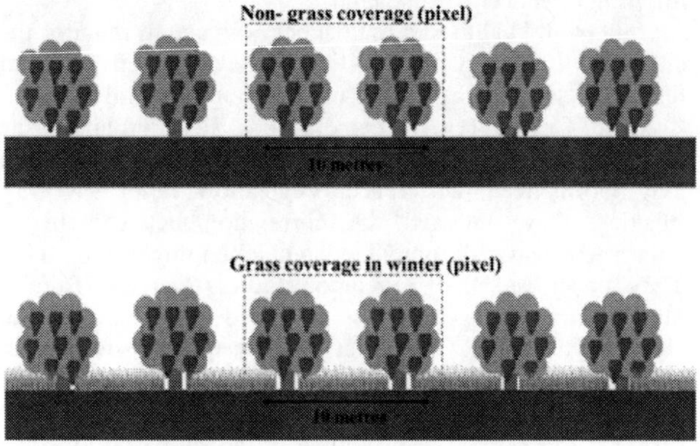


Figure 1. Scheme of pixel information content.

TABLE II. SUMMARY OF EXPECTED CHANGES ACCORDING TO OUR HYPOTHESIS

| Bands | Reflectance GC=1 | Reflectance GC=0 | Differences in reflectance GC=1 | Differences in reflectance GC=0 |
|---|---|---|---|---|
| B2 | Low | Low | Low | Low |
| B3 | Higher | High | High | Low |
| B4 | Low | High | High | Low |
| B8 | High | High | Low | Low |
| B9 | Higher | High | High | Low |
| **Pixels of:** | **GC=1 Winter** | **GC=1 Summer** | **GC=0 Winter** | **GC=1 Summer** |
| Vid | High percentage | High percentage | High percentage | High percentage |
| Soil | Almost null | Low percentage | Low percentage | Almost null |
| Green grass coverage | Low percentage | Almost null | Almost null | Almost null |

Figure 2. Classification of plots in the studied zone.

In this context, we use some operations using this software. The first operation that we apply is the "Raster calculator". This tool allows us to observe the qualitative differences between the same areas on both dates. The output for these tools is a new raster in which the pixel values are the differences between the initial (January) and final (June) scenarios. Thus, we will be able to identify the different colouration in pixels that represent plots with or without grass coverage. In addition, we will extract quantitative information as a raster by applying "Zonal Statistics as table" to obtain a table with pixel values for each plot.

Finally, with the obtained data, we will use statistical software, Statgraphics Centurion XVIII [17]. With this software, we will perform statistical analyses to determine if the observed differences are statistically significant.

## IV. RESULTS

In this section, we display the obtained images and the time series analysis that we use to determine the existence of grass coverage. First, we evaluate the different bands, analyzing which combination provides the highest visual difference between plots with and without grass coverage. Finally, we use statistical analysis to verify if differences are statistically significant or not.

### A. *Band combination*

According to the analysis of data of obtained images, we can identify the following differences. In the range of bands from the visible spectrum (B2 to B4) of January, we find that the green band has the highest pixel values in plots with grass coverage. The blue band presents medium values, and the red band has the minimum pixel values in the plots with the grass coverage. On the other hand, in non-covered plots, the red band obtains higher values because there is more soil represented in the pixel, and it increases the reflectance in that specific wavelength. In addition, between NIR, WVP, and NDVI bands for the same season, the NIR band presents the highest pixel values for covered and no-covered plots. Figure 3 represents the RGB composition band and the single bands displayed. Besides, Figure 4 displays the other bands: the NIR band, WVP, and NDVI band.

In order to evaluate the differences in grass coverage in all the selected plots, it is necessary to combine each one of the bands from winter and summer seasons as pointed in the time series analysis. Figure 5 shows the results of the combination of bands. In this case, red, green, blue, NIR, WVP, and NDVI seasonal band combinations displayed that the plots with grass coverage present low pixel values, represented in a darker colour. The pixel represents the difference between the winter and summer values in each band. Besides, plots without grass coverage will display higher pixel values, symbolized in a lighter colour.

Figure 5 displays the image combination results. Moreover, in the visible spectrum (red, green, and blue bands), the obtained changes are not visually significant. Nevertheless, in the NIR, WVP, and NDVI bands, we can observe differences in pixel values between the selected plots (red plots and blue plots).

Considering the spatial resolution of used images, it is relativity challenging to observe visual changes in the crops. Therefore, the use of a statistical analysis will be necessary to determine which band combination is the best one for detecting the presence of grass.

### B. *Statistical analysis*

It is necessary to quantify the changes in the pixel value numerically. The Statgraphics software allows us to analyze the pixel values obtained from the combined results, getting statistical information from each image.

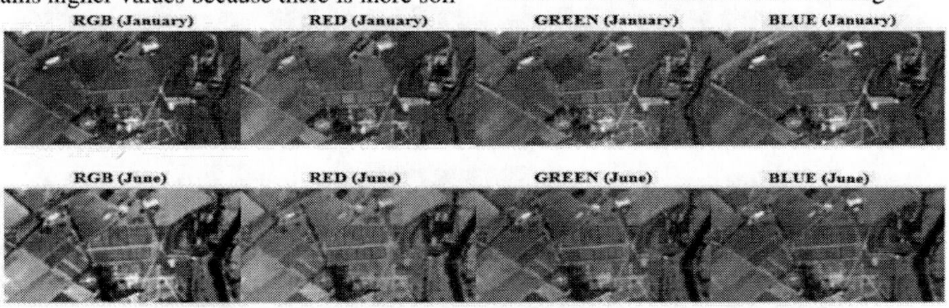


Figure 3. Visible spectrum bands.

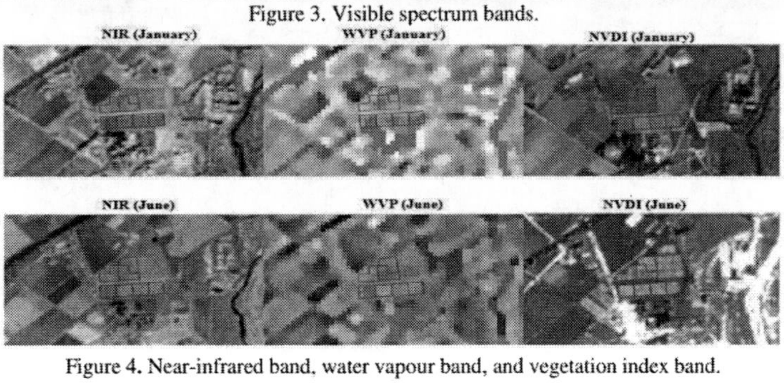


Figure 4. Near-infrared band, water vapour band, and vegetation index band.

Table 3 displays the "MEAN" for all the bands for different value in each one of the evaluated plots. With this, we can observe the differences between winter and summer, analyzing which band is the best for detecting changes.

Then, the application of an Analysis of Variance (ANOVA) procedure is required. This method is used to evaluate the values for each band combination and determine the significate grade of the values. To better illustrate the differences between the plots with and without grass coverage, we present Figure 6. Figure 6 is composed of six graphics. These graphics represent the data of the two classifications of plots as a box and whiskers diagrams. The box and whiskers diagrams, also known as box-plots, represent the similarity between the values of grass-covered plots and plots without grass. The data present higher similarities among them when there are higher overlapping between the plots classified as 1 and 0. The purpose is to find the band for which the difference among the calculated raster is maximum for both groups (GC=1 and GC=0). In this case, we can observe that all graphics indicate a certain distance between the values of both groups of plots. Nonetheless, the blue band is the only one that has the values of both groups close to each other. To complete the verification, we determinate the statistical significate of the values. We use the p-value to verify the significance of the observed differences. To be considered as a significant difference, the p-value must be smaller than 0.05.

Table 4 summarizes de p-values for all the band combinations. We observe that all the bands have significant values, except for the blue band. The best range of values is represented in the NIR, NDVI, and WVP bands, with the most accurate results being of the NIR band (p-value of 0.0020).

TABLE III. MEAN OF THE PIXEL VALUES OF DIFFERENT BAND COMBINATIONS FOR COVERED AND NON-COVERED PLOTS.

| Classific. | Red | Green | Blue | NIR | WVP | NDVI |
|---|---|---|---|---|---|---|
| 1.GC=0 | -775 | -632 | -456 | -1336 | -831 | -0,12 |
| 2. GC=0 | -618 | -623 | -422 | -1464 | -881 | -0,16 |
| 3. GC=0 | -775 | -697 | -491 | -1399 | -871 | -0,11 |
| 4. GC=0 | -892 | -756 | -557 | -1411 | -870 | -0,09 |
| 5. GC=1 | -1246 | -854 | -620 | -1113 | -785 | 0,02 |
| 6. GC=1 | -1179 | -797 | -582 | -1006 | -763 | 0,03 |
| 7. GC=1 | -1061 | -798 | -553 | -1176 | -799 | -0,02 |

Figure 5. Results of combined images of January and June.

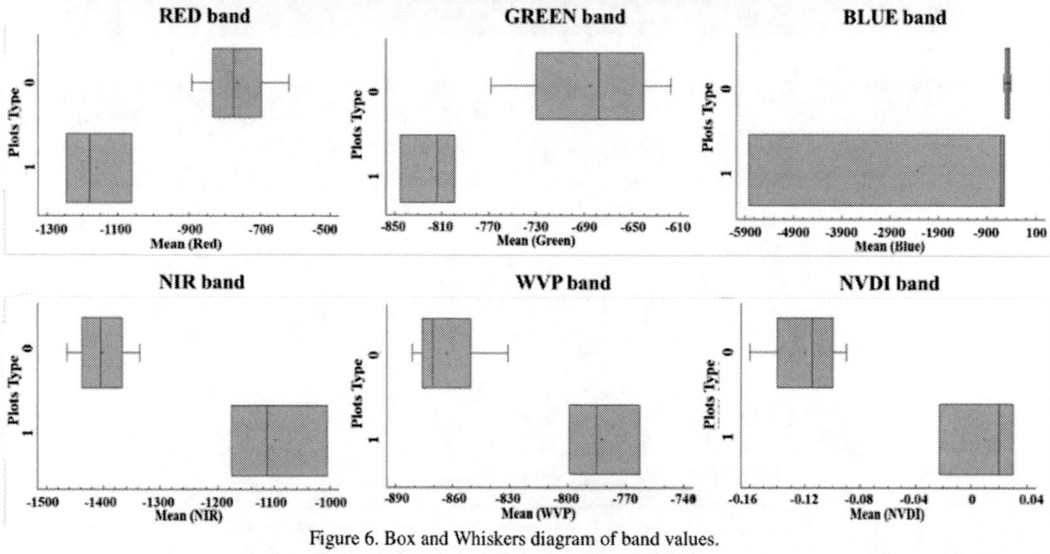


Figure 6. Box and Whiskers diagram of band values.

TABLE IV. THE P-VALUE OF ANOVA ANALYSIS FOR SELECTED BANDS

| Bands | B4 | B3 | B2 | B8 | B9 | NDVI |
|---|---|---|---|---|---|---|
| p-Value | 0.0043 | 0.0184 | 0.2611 | 0.0020 | 0.0036 | 0.0021 |

The results show that the best band to differentiate whether vineyard crops have grass cover or not is the NIR band. The pixel values of the resultant raster (time series analysis) for this band have different values for plots wit and without grass coverage. Plots with grass coverage have pixel values from -1000 to -1200. On the other hand, plots without grass coverage are composed of pixels with values from -1200 to -1500.

Although we have demonstrated that the time series analysis might help to overcome the drawback of the low spatial resolution in this application, the methodology used in this study presents certain limitations. The presence of clouds might be a limitation in cases that, due to life cycles of crop and grass, images form periods with high presence of clouds are required. Moreover, the technique used could increase its efficiency considerably in plots with a greater extension, resulting in getting relevant information. Furthermore, due to the limited extension of the studied area and its low variability, our results cannot be applied in the different regions of Spain.

## V. CONCLUSION

We present a methodology, based on time series analysis to determine the existence or absence of grass coverage in vineyards. The target of this system is to provide a fast, remote, easy to use, and cheap method to evaluate the adoption of conservation agriculture. In this study, we use Sentinel-2 images, using different bands such as red, green, blue, NIR, WVP, and NDVI bands. The comparison of bands from different moments of the year allows us to evaluate the changes between the selected plots to determine the existence or absence of grass coverage. Our results indicated that the best band for the time series analysis is NIR band, followed by the NDVI band, and WVP band.

For future work, we will improve our study by including different region vineyards to verify the efficiency of this application. Furthermore, we will evaluate if it is possible to use other time series combinations avoiding the use of data from January due to the high probability of clouds in this month. Finally, we plan to test this method in orange crops. These present an added difficulty due to the position of the crops (trees more closely together) and the arrangement of leaves throughout the year, which may cause a greater probability of error when detecting the grass cover. The introduction of a soil gloss correction, such as the SAVI and MSAVI, will also be explored.

## ACKNOWLEDGEMENT

This work has been partially funded by the European Union through the ERANETMED (project ERANETMED3-227 SMARTWATIR, by the "Ministerio de Economía y Competitividad" in the "Programa Estatal de Fomento de la Investigación Científica y Técnica de Excelencia, Subprograma Estatal de Generación de Conocimiento" within the project under Grant TIN2017-84802-C2-1-P, and by Conselleria de Educación, Cultura y Deporte with the Subvenciones para la contratación de personal investigador en fase postdoctoral, grant number APOSTD/2019/04.

## REFERENCES

[1] O. F. Godber and R. Wall, "Livestock and food security: vulnerability to population growth and climate change", Global change biology, vol. 20, no 10, pp. 3092-3102, 2014.

[2] S. J. Vermeulen, B. M. Campbell, and J. S. I. Ingram, "Climate change and food systems", Annual review of environment and resources, vol. 37, pp. 195-222, 2012.

[3] M. Ruiz-Colmenero, R. Bienes, and M. J. Marques, "Soil and water conservation dilemmas associated with the use of green cover in steep vineyards", Soil and Tillage Research, vol. 117, pp. 211-223, 2011.

[4] R. Gebbers and V. I. Adamchuk, "Precision agriculture and food security", Science, vol. 327, no 5967, pp. 828-831. 2010

[5] C. Thierfelder, S. Cheesman, and L. Rusinamhodzi, "A comparative analysis of conservation agriculture systems: Benefits and challenges of rotations and intercropping in Zimbabwe", Field crops research, vol. 137, pp. 237-250, 2012.

[6] L. Parra et al., "Edge detection for weed recognition in lawns", *Computers and Electronics in Agriculture*, 2020, vol. 176, p. 105684.

[7] J. K. Patil and R. Kumar, "Advances in image processing for detection of plant diseases", *Journal of Advanced Bioinformatics Applications and Research*, vol. 2, no 2, pp. 135-141, 2011.

[8] L. Garcia et al., "Quantifying the Production of Fruit-Bearing Trees Using Image Processing Techniques", *In The Eighth International Conference on Communications, Computation, Networks, and Technologies (INNOV19)*, pp. 14-19, 2019.

[9] M. Possoch et al., "Multi-temporal crop surface models combined with the RGB vegetation index from UAV-based images for forage monitoring in grassland", *The International Archives of Photogrammetry, Remote Sensing and Spatial Information Sciences*, vol. 41, pp. 991, 2016.

[10] F. A. Vega, F. C. Ramirez, M .P. Saiz, and F. O. Rosua, "Multi-temporal imaging using an unmanned aerial vehicle for monitoring a sunflower crop", Biosystems Engineering, vol. 132, pp. 19-27, 2015.

[11] H. Sun et al., "Monitoring of maize chlorophyll content based on multispectral vegetation indice", Multispectral, Hyperspectral, and Ultraspectral Remote Sensing Technology, Techniques and Applications IV. International Society for Optics and Photonics, 2012. pp. 852711.

[12] N. B. Mishra and K. A. Crews, "Mapping vegetation morphology types in a dry savanna ecosystem: integrating hierarchical object-based image analysis with Random Forest", Int. Journal of Remote Sensing, vol. 35, no 3, pp. 1175-1198, 2014.

[13] L. Parra et al., "Comparison of Single Image Processing Techniques and Their Combination for Detection of Weed in Lawns", International Journal On Advances in Intelligent Systems, Valencia, 2019, vol.12, no. 3–4, pp. 177–190.

[14] J. G. Clevers and A. A. Gitelson, "Remote estimation of crop and grass chlorophyll and nitrogen content using red-edge bands on Sentinel-2 and-3", International Journal of Applied Earth Observation and Geoinformation, vol. 23, pp. 344-351, 2013.

[15] C. A. Rokhmana, "The potential of UAV-based remote sensing for supporting precision agriculture in Indonesia", Procedia Environmental Sciences, vol. 24, no 2015, pp. 245-253, 2015.

[16] Esri. ArcGIS. [Online]. Available from: https://www.esri.es/es-es/home [Retrieved: October, 2020].

[17] Statgraphics. [Online]. Available from: https://statgraphics.net/. [Retrieved: October, 2020].

**International Academy, Research, and Industry Association (IARIA)**
412 Derby Way
Wilmington, DE 19810

ISBN 978-1-7138-2178-6

CURRAN ASSOCIATES INC.
proceedings.com

# Thirteenth International Conference on Mobile Ubiquitous Computing, Systems, Services and Technologies (UBICOMM 2019)

Held at NexTech 2019

Porto, Portugal
22 – 26 September 2019

IARIA